Fundamentals of

Groundwater and Drainage Engineering

The Authors

Dr. A. K. Vashisht is presently working as Associate Professor in the Department of Irrigation and Drainage Engineering, College of Agricultural Engineering and Post-Harvest Technology, Central Agricultural University (Imphal), Ranipool, Gangtok, Sikkim. Before joining to the present position, he served as Junior Research Officer/Assistant Engineer (Design) in the Department of Irrigation and Drainage Engineering, G. B. Pant University of Agriculture and Technology, Pantnagar, Uttarakhand for four years from February 2006 to January 2010. He was co-PI of All India Coordinated Research Project on "Groundwater Utilization" (now named as "Irrigation Water Management"). During his service in three premier institutions of India (i.e. PAU, Ludhiana; GBPUAT, Pantnagar; CAU, Imphal), he got exposure to the problems like declining water-table and brackish groundwater of Punjab, scarcity of water in *Shiwalik* foothills, drying-up of artesian wells in *Tarai* region and springs in lower and mid-Himalayan regions of Uttarakhand and Sikkim. For tackling with the mentioned issues, he proposed various management/adaptive strategies through research projects and publications. In his regular service experience of more than ten years, he has authored 51 research publications which include international/national research papers, papers in conference proceedings, book chapters, popular articles, and laboratory manuals. At present, he is PI of one externally funded (DST, Govt. of India) and one intramural research project (CAU, Imphal). Before joining the GBPUAT, Pantnagar, he worked as Post-doctorate research fellow for about two years in the Department of Soil and Water Engineering, Punjab Agricultural University, Ludhiana. Earlier, he obtained his M. Tech. and Ph. D. degrees from Punjab Agricultural University, Ludhiana with specializations in well hydraulics and brackish groundwater management, respectively.

Dr. Rajan Aggarwal is presently working as Senior Research Engineer and Head of Department of Soil and Water Engineering, Punjab Agricultural University, Ludhiana, Punjab. He is also PI of All India Coordinated Research Project on "Irrigation Water Management" and Project on "Groundwater Resource Management to Mitigate the Impact of Climate Change in Punjab and Haryana (NICRA)". He has more than 25 years of teaching, research, and extension experience. In addition, he has already completed seven research projects as Co-PI/Collaborator. For managing existing water resources of the Punjab, he is involved in the research on groundwater modeling, rainwater harvesting, climate change, pump testing, and drip irrigation. He has designed and installed more than hundred rainwater harvesting structures at different sites in the state. He has more than 160 publications in national/international journals, popular articles, book chapters etc. He has guided several B. Tech./M. Tech., and Ph. D. students in his professional career.

Fundamentals of

Groundwater and Drainage Engineering

— *Authors* —

Dr. A K Vashisht

Dr. Rajan Aggarwal

2018

Daya Publishing House®

A Division of

Astral International Pvt. Ltd.

New Delhi – 110 002

ISBN 9789388173025 (Int. Edition)

Publisher's Note:

Every possible effort has been made to ensure that the information contained in this book is accurate at the time of going to press, and the publisher and author cannot accept responsibility for any errors or omissions, however caused. No responsibility for loss or damage occasioned to any person acting, or refraining from action, as a result of the material in this publication can be accepted by the editor, the publisher or the author. The Publisher is not associated with any product or vendor mentioned in the book. The contents of this work are intended to further general scientific research, understanding and discussion only. Readers should consult with a specialist where appropriate.

Every effort has been made to trace the owners of copyright material used in this book, if any. The author and the publisher will be grateful for any omission brought to their notice for acknowledgement in the future editions of the book.

Published by : **Daya Publishing House®**
A Division of
Astral International Pvt. Ltd.
– ISO 9001:2015 Certified Company –
4736/23, Ansari Road, Darya Ganj
New Delhi-110 002
Ph. 011-43549197, 23278134
E-mail: info@astralint.com
Website: www.astralint.com

Digitally Printed at : **Replika Press Pvt. Ltd.**

Dedicated to...

...my grandfather late **Sh. Harbans Lal Vashisht**

— *A. K. Vashisht*

...my beloved wife **Mrs. Amita Aggarwal**

— *Rajan Aggarwal*

Preface

This book is written to serve as a ready reference for students of agricultural and civil engineering degree programs. It will also serve as a text book for the students at the undergraduate level in Agricultural Engineering/Agricultural colleges. The contents of the book are strictly in accordance with the syllabus of fifth Deans' committee recommendations. Most importantly, this book is a valuable reference for the students appearing in the various competitive examinations. The book is meant for the students for excelling in any competitive examination through strong basic fundamentals. Therefore, the authors have made an effort to present fundamentals related to the subject in concise form. As far as possible, the theory has been explained through schematic diagrams and photographs. Because, the book is in concise form, students can refresh their basic fundamentals within limited available time before examinations.

Fundamentals of Groundwater and Drainage Engineering, is a comprehensive treatment on groundwater, well hydraulics, well construction, water lifting devices, waterlogged soils, soil permeability measurement, equations for evaluating depth and spacing of drains, surface and subsurface drains, tube well drainage system, and salt affected soils. This book also contains selective objective type questions.

While writing the book, it has been assured by the authors that the theory presented in the book should be technically correct. The figures, tables and articles are numbered chapter-wise. To ease the search of relevant topic/term in the book, an elaborated index has also been provided in the end.

The authors are thankful to all their students, whose typical queries regarding the subject motivated us to write this manuscript. The authors are grateful to their teachers specifically, Dr. S. D. Khepar, Dr. S. K. Sondhi, Dr. N. K. Narda, Dr. S. K. Shakya, Dr. D. S. Taneja, Dr. K. G. Singh, Dr. Tejwant Singh, and Dr. D. S. Pal, whose able guidance has enthused us to achieve this aim. Authors also wish to acknowledge the All India Coordinated Research Project (Indian Council of Agricultural Research, New Delhi) on Irrigation Water Management (formerly known as Groundwater Utilization) of two centers *viz*. Punjab Agricultural University, Ludhiana and G. B. Pant University of Agriculture and Technology, Pantnagar as the field experience gained by us while working in this project assisted us to create/present the more realistic picture of the topics related to subject. Following the ethics, the authors have acknowledged the sources of information. In spite of this effort, any omission therein is inadvertent. The authors welcome the suggestions from the students and teachers to improve the text.

A. K. Vashisht

R. Aggarwal

Contents

1 Groundwater

1.1 Introduction

That water which occupies all the voids within a geologic stratum is referred to as groundwater. It can be distinguished from the unsaturated zone in the sense that where voids are filled with water and air. Moreover, the unsaturated zones are usually found above the saturated zones. It is rather difficult to demarcate the waters between two zones. The need for freshwater boosts the humankind to invent various methods to develop groundwater. Groundwater is an important component of earth's water circulatory system (popularly known as hydrologic cycle). After entering through the recharging zones (i.e. ground surface, water bodies), water enters the underground porous formations. These formations act as conduits for transmission and as reservoirs for storage of water. The extensive storage capacity of the formations along with the slow velocity of flow makes the groundwater a reliable source of water.

The origin of all groundwater is surface water. Precipitation, stream-flow, lakes, and natural ponds are the main sources from where recharging to groundwater takes place. After infiltrating through the ground surface, water percolates vertically through the unsaturated medium under the force of gravity whereas, on meeting the saturated zone its direction of movement is guided by the hydraulic gradient at the location. Pumping through wells is the largest mean of withdrawing groundwater artificially.

In this chapter, we will deal with the definitions related to groundwater, dynamic groundwater resources of India, groundwater investigations, Darcy's law, and various aquifer properties.

1.2 Definitions

Groundwater – The term *groundwater* is used to denote all the waters found beneath the ground surface.

Aquifer – An *aquifer* is a geologic formation (or a group of formations) saturated with water and which permits its withdrawal in significant amount and rate. The word is from Latin origin. The prefix '*Aqui-*' comes from '*aqua*' which means "water" and suffix '*–fer*' comes from '*ferre*' which means "to bear". Hence, the complete meaning of aquifer is a *water bearer*. Coarse sand acts as a good aquifer material. The formations like alluvial deposits, limestone, volcanic rock, sandstone, igneous and metamorphic rocks, and clayey formations act as good aquifers. About 90% of all developed aquifers around the world are alluvial deposits. In India, major alluvial aquifers are quaternary rock formations, which comprise recent, older, Aeolian, and coastal alluvium and essentially composed of kankar, pebbles, sand, silt, and clay.

Aquitard – An *aquitard* is a saturated geologic formation having poor permeability and thus have a very limited capacity of transmitting water through it (Figure 1.1). The suffix '*-tard*' comes from the Latin word '*tardus*' which means "*slow*". More commonly, aquitard is referred as *semi-pervious formation* or *leaky formation*. Sandy clay layer is an example of aquitard.

Aquiclude – An *aquiclude* is a geologic formation, which may contain water in appreciable quantities, but is unable to transmit it in significant quantities. The suffix '*-clude*' comes from the Latin word '*claudere*' which means, "*to close*". Clay layer acts as an aquiclude.

Aquifuge – An *aquifuge* is a geologic formation, which neither contains nor transmits water. The suffix '*-fuge*' comes from the Latin word '*fugere*' which means, "*to drive away*". Solid granite is an example of aquifuge.

Piezometric surface – An imaginary surface defined by the water levels in a number of piezometers tapping a confined/semi-confined aquifer is called the *piezometric surface* (or *isopiestic surface*).

Confined aquifer – An aquifer which is bounded from above and below by impervious formations is termed as *confined aquifer* and is also known as *pressure* or *artesian* aquifer (Figure 1.1).

Semi-confined aquifer – An aquifer that is bounded from above by a semi-pervious layer and below by a semi-pervious or impervious is known as *semi-confined* or *leaky* aquifer. Before pumping of leaky aquifer, its piezometric level matches with the water table of upper phreatic aquifer. During the pumping, the simultaneous lowering of phreatic and piezometric levels describe that aquifer may be termed as *leaky aquifer with prompt yield*. However, when there is a significant lag between the lowering of two water levels, the aquifer may be termed as *leaky aquifer with delayed yield*.

Unconfined aquifer – An *unconfined aquifer* (also called *phreatic aquifer, water table aquifer*) is that in which water table acts as its upper boundary (Figure 1.1). A phreatic aquifer resting on a leaky layer is termed as *leaky phreatic aquifer*.

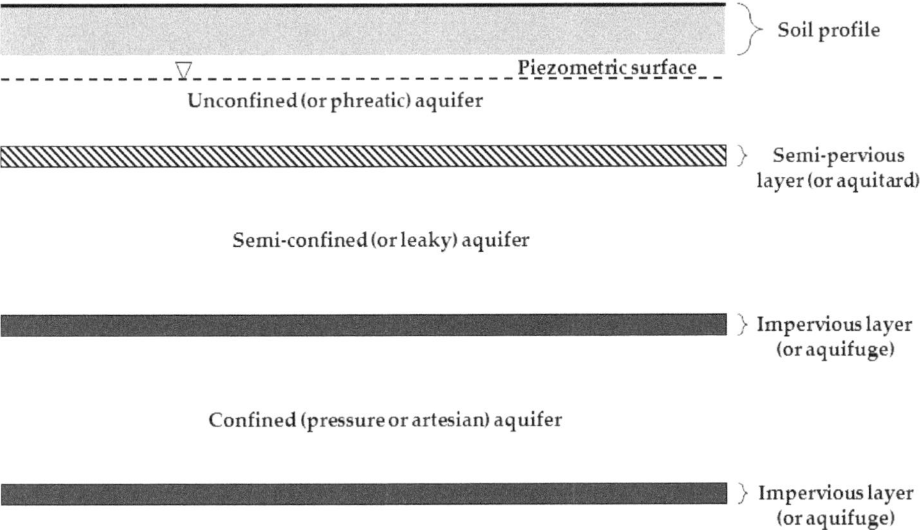

Figure 1.1: Definition sketch of different types of confining layers and aquifers.

Perched aquifer – The occurrence of an impervious/semi-pervious layer of limited areal extent between the water table and the ground surface leads to the existence of a small phreatic aquifer, which is termed as *perched aquifer* (Figure 1.2). Clay or loam lenses in sedimentary deposits often have shallow perched aquifers. The upper surface of the groundwater in perched aquifer is known as *perched water table*.

Idealized aquifer – A geologic formation having hydrologic properties identical everywhere is recognized as *idealized aquifer*. When the properties of the aquifer are independent of direction, it is known as *isotropic aquifer*.

1.3 Groundwater dynamics

The main source of groundwater recharging is rainfall, which annually replenishes the groundwater resources. Consequently, the nature of groundwater resources becomes dynamic. About 120 cm depth of rainfall is received annually in of which 75% occurs in the monsoon period (i.e. June to September). For studying the hydrogeological characteristics of the underground formations, monitoring of water level is the basic data element. All over India, central ground water board has installed 15640 monitoring wells for observing the water level and its quality (CGWB, 2011). The data is procured in the months of January, April/May, August, and November.

The annual replenish-able groundwater resources for the India have been assessed as 433 billion cubic meter (BCM) (CGWB, 2011). Monsoon rainfall contributes 253 BCM of water through recharging, whereas rainfall received during the months other than monsoon months (i.e. non-monsoon) contributes 41 BCM. The rest 139 BCM of water is recharged through canal seepage, return flow from

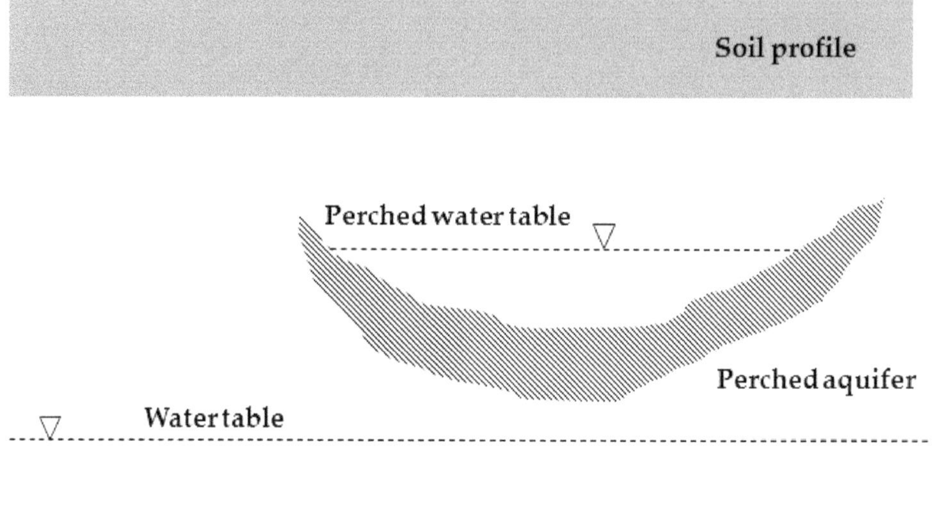

Figure 1.2: Definition sketch of a perched aquifer.

irrigation, recharge from tanks, ponds, and other water conservation structures. For the year 2010-11, the annual groundwater draft of the whole country was estimated equal to 245 BCM. Irrigation sector is consuming 91% of the total annual groundwater draft (i.e. 223 BCM).

1.4 Artificial groundwater recharging

Artificial recharge may be defined as boosting the natural movement of surface water into underground formations by some method of construction or by directly injecting it into aquifer. The selection of particular method for artificial groundwater recharge is affected by (i) topography of area, (ii) type of soil, (iii) general land slope, (iv) depth to impermeable layer, (v) land value, (vi) water quality, and (vii) quantity of water to be recharged. Water spreading is the most widely used practice for recharging groundwater artificially. Depending upon the way the water is conveyed and allowed to store on the ground surface, the spreading methods are further classified as *basin, flooding, ditch* and *furrow, stream-channel, irrigation,* and *water conveyance network.* Basins can be constructed by making dikes or levees on the normal land surface by transporting soil from other place. Usually, a series of basins are constructed along the natural stream (Figure 1.3). In this method, around

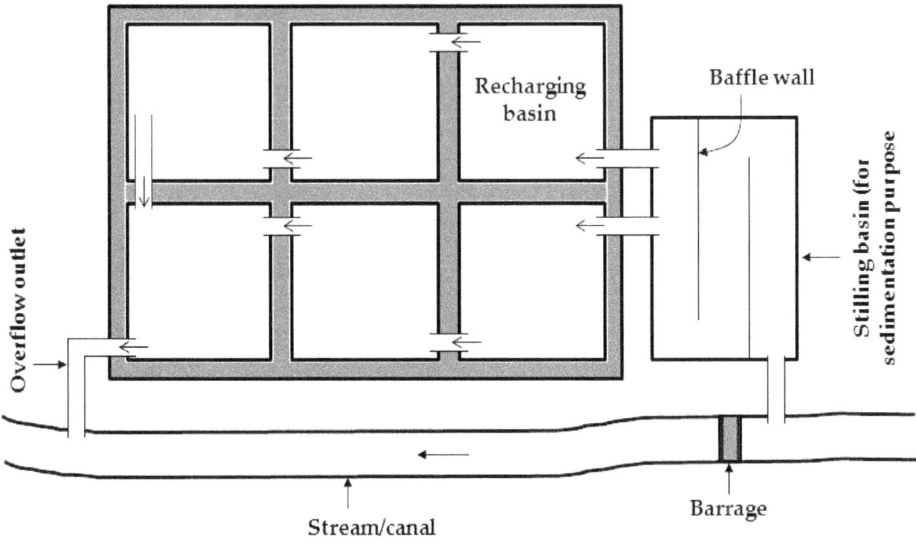

Figure 1.3: Schematic sketch of the multiple-basin recharging method.

75-90% of the gross area of the project is under water contact.

Spreading of water evenly in the form of a thin sheet over a large area is termed as *flooding*. Water is released on the surface with minimum velocity to avoid disturbing the soil. In ditch and furrow method, water is distributed to a series of ditches, or furrows, that are shallow, flat-bottomed, and closely spaced. The width of the ditch is generally kept in the range of 0.3 to 1.8 m.

Natural streams with little modification can be used to enhance groundwater recharging. The concept is to increase the residence time as well as surface area contact of water on the stream surface. Recharging through unlined canals is a very effective method of groundwater recharging in the areas having dense canal network. The construction of low-height dikes in the canals increases the contact time with the seepage surface (Figure 1.4a). To further augment the recharge rate, shallow depth shafts/wells may be constructed in the canal bed at regular intervals (Figure 1.4b).

Depending upon the depth of this impermeable layer, groundwater can be recharged using pits or wells. Wherever the impermeable layers are at shallow depth, pits can be excavated up to permeable formations. Smaller depth pits are usually filled with gravels up to the top. In recharge well method, surface water after due filtration is directly admitted to the freshwater aquifers. During recharging, the flow through well will be in reverse direction and because of it cone of impression is formed instead of cone of depression (Figure 1.5).

Aquifer-storage-recovery (ASR) is a technique in which good/potable quality water is stored in the aquifer having poor quality water. When required, the stored water is pumped later on. Vashisht and Shakya (2016a-b) have successfully demonstrated the storage of good quality water in brackish aquifer. The

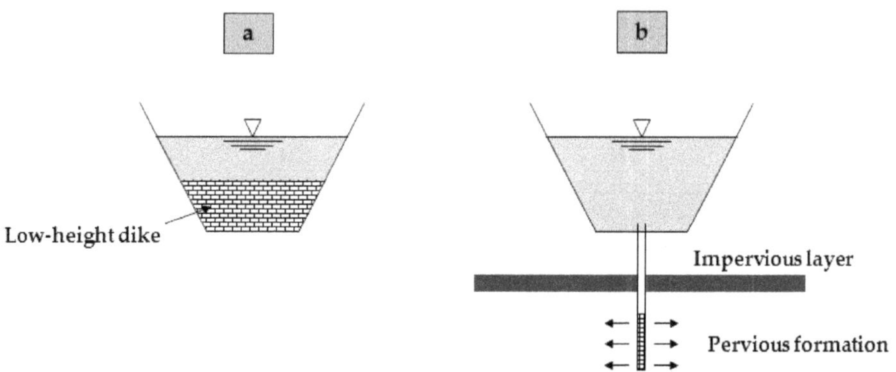

Figure 1.4: Schematic sketches of (a) low-height dike in canal bed, and (b) shallow depth shaft/well.

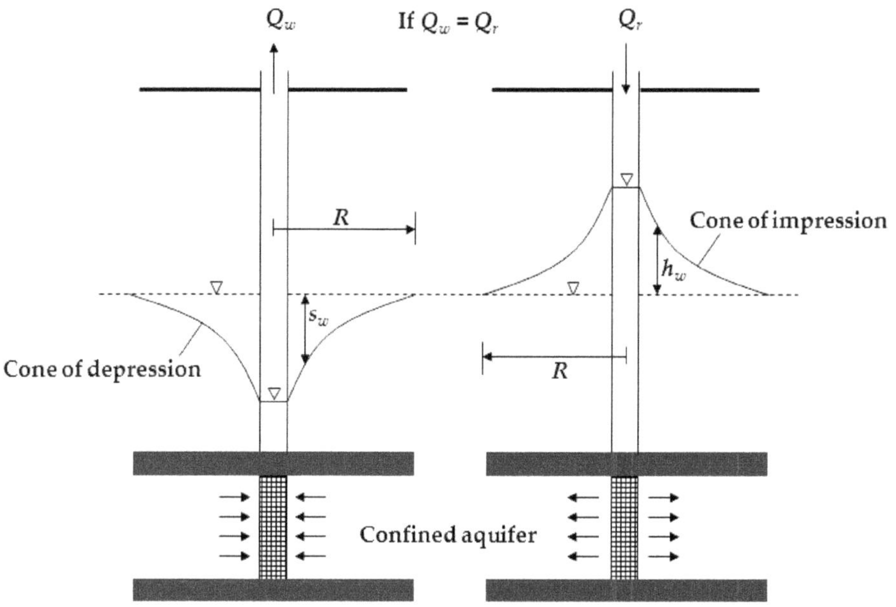

Figure 1.5: Schematic sketch showing the difference of a recharge well from the discharge well.

demonstrated technique was altogether different from the previous ASR technique (in which freshwater is stored in the brackish water) in the sense that water in the proposed technique is stored at top of the brackish water layer with the aid of lesser capacity multiple well points system.

In induced recharge method, groundwater is pumped adjacent to a river or lake so that the lowering of groundwater level with respect to the level of surface water source will create a hydraulic gradient towards well pumping zone. This method is most suitable where the natural streams or other surface bodies are in hydraulic contact with aquifer through permeable media.

1.5 Subsurface investigations

The purpose of *test drilling* is to explore the groundwater resources of a region and to obtain information on the geologic formations encountered depth-wise. The complete information when depicted depth-wise is termed as *well log*. *Electrical resistivity method* of groundwater exploration is based on the measurement of electrical resistivity of underground formations that vary with certain aquifer characteristics. A clean sand formation saturated with freshwater shows high resistivity and is considered as potential aquifer. *Electrical resistivity logging* is affected by the chemical quality of the groundwater. Most importantly, it can be measured only in mud filled, uncased boreholes. *Gamma-ray logging* is based on the measurement of gamma radiation (in counts per second) emitted by subsurface formations. In comparison to electrical resistivity logging, gamma-ray logging is feasible in both cased and un-cased boreholes. Moreover, it is also not affected by the change in groundwater quality. *Electrical resistivity surveying* is performed at the ground surface using four electrodes that are arranged at equal distances in a straight line. Outer electrodes are used to apply current to the ground and the inner are used to measure potential drop. *Seismic refraction surveying* is based on the travelling velocities of shock waves through different geological formations. Shock waves travel faster in denser formations.

1.6 Darcy's law

Based on the experimental results, Darcy (1856) concluded that the rate of flow Q through a vertical homogeneous sand column is (i) proportional to the column's cross-sectional area (A), (ii) proportional to the difference of hydraulic heads at the two ends $(h_1 - h_2) = h_L$, and (iii) inversely proportional to the column length (L) (Figure 1.6). Then the equation for evaluating flow (Q) through the porous media will be

$$Q \propto vA \quad or \quad Q \propto \frac{h_L}{L} A \tag{1.1}$$

Incorporating the proportionality sign in Eq. (1.1), *Darcy formula* can be written as

$$Q = -K\frac{(h_1 - h_2)}{L} A \Rightarrow -K\,iA \quad or \quad v \Rightarrow \frac{Q}{A} = -K\,i \quad or \quad K = -\frac{v}{i} \tag{1.2}$$

In Eq.s (1.1-1.2), v is the *specific discharge* or *Darcy velocity* and is defined as flow per unit area; K is the coefficient of proportionality and is termed as *hydraulic conductivity* of the sand column and may be defined as specific discharge per unit hydraulic gradient. The negative sign in Darcy's law indicates that the flow of water is in the direction of decreasing head. During the evaluation of Darcy velocity, entire cross-sectional area of the aquifer is considered without regard to solids and pores.

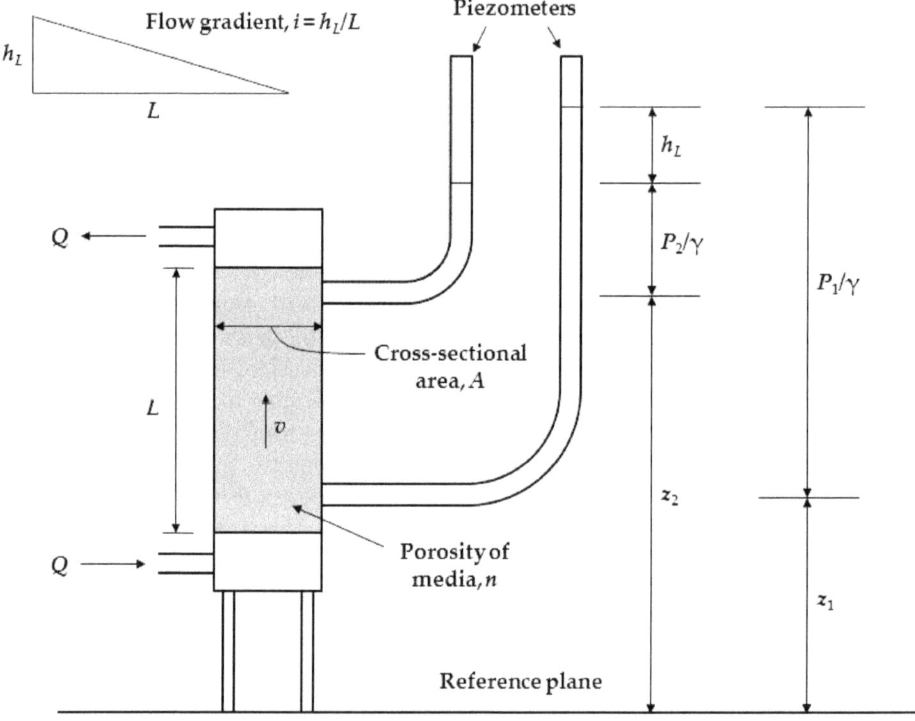

Figure 1.6: Schematic sketch showing the experimental verification of Darcy's law.

However, the groundwater flow through aquifer follows the tortuous pore space only. Hence, the *average interstitial velocity* through the pores can be evaluated as

$$v_n = \frac{Q}{nA} \quad or \quad v_n = \frac{v}{n} \tag{1.3}$$

In Eq. (1.3), n refers to the porosity of aquifer. For example, if porosity of the aquifer is 50%, then according to Eq. (1.3), the interstitial velocity will be equal to $2v$.

1.7 Aquifer properties

1.7.1 *Permeability* – *Permeability* is the ability of a porous medium to transmit a fluid. It is a medium property and is independent of fluid properties. Mathematically, it can be expressed as

$$k = \frac{K\mu}{\rho g} \tag{1.4}$$

In Eq. (1.4), K is the hydraulic conductivity; μ is the dynamic viscosity; ρ is the fluid density; and g is the acceleration due to gravity. The permeability has units of area. The value of permeability (k) in petroleum industry is measured in the unit named *darcy* and 1 *darcy* = 0.987 $(\mu m)^2$.

1.7.2 *Transmissivity* (T) – Whenever flow through the entire thickness of aquifer is considered, the term aquifer transmissivity comes into picture. *Transmissivity* is an *aquifer characteristic*, which may be defined as the rate of flow through a vertical strip of the aquifer of unit width and extending throughout the full saturated thickness of the aquifer. Most importantly, the concept is valid only in two-dimensional flow (or flow towards well fully penetrating the aquifer). For satisfactory performance of irrigation well, the transmissivity value of the aquifer must not be less than 120 m^2/d.

1.7.3 *Storage coefficient* (S) – Pumping/recharging of an aquifer represents the change in storage volume of the aquifer. In confined aquifers, weight of the overburden is mainly supported by the solid structure of the aquifer and partially by the hydrostatic pressure. Reduction in hydrostatic pressure (such as due to pumping) transfers the weight to solid structure which compresses the aquifer. This compression load forces the aquifer to release some water from it. Hence, Storage coefficient is defined as the volume of water a confined aquifer releases from or takes into storage per unit surface area of the aquifer per unit change in component of *piezometric head* normal to surface. It is a property related to confined and semi-confined aquifers and is a dimensionless quantity. In general, the value of S lies in the range from 5×10^{-5} to 5×10^{-3}. According to Lohman (1972), storage coefficient varies with the aquifer thickness as per relation

$$S = 3\times10^{-6} b \tag{1.5}$$

In Eq. (1.5), b is the aquifer thickness in meters.

1.7.4 *Specific yield* – Specific yield is defined as the volume of water stored or releases per unit surface area of aquifer per unit change in the *water table* normal to that surface. In other terms, it is the ratio of volume of water that can be drained (after saturation) by gravity to its own volume. Specific yield depends on grain size, shape and distribution of pores, compaction of the stratum, and time of drainage. It is a property of an unconfined aquifer. Specific yield is the quantity of water that a unit volume of aquifer material will give up when drained under gravity. The quantity of water that a unit volume of aquifer retains when subjected to gravity drainage is called *specific retention*. It is worth to mention here that the term *specific retention* is synonymous with *field capacity*. The *porosity* of an aquifer is sum of its specific yield and specific retention. Specific yield can be measured by laboratory, field, and empirical techniques but the methods based on the pumping tests are more reliable.

1.7.5 *Hydraulic resistance* (c) – It is a property of semi-pervious or leaky confining layer. It defines the resistance caused by the layer to *vertical flow* of water through it. It is evaluated by dividing saturated semi-pervious layer's thickness (b') by its vertical hydraulic conductivity (K') as

$$c = \frac{b'}{K'} \tag{1.6}$$

The hydraulic resistance has units of time. A geologic formation with infinite hydraulic resistance is termed as impervious layer (as $K' \to 0, \quad c \to \infty$).

1.7.6 *Leakage factor* (B) – It is a property of leaky aquifer and is evaluated using relation

$$B = \sqrt{Kb\,c} \tag{1.7}$$

In Eq. (1.7), K and b are the hydraulic conductivity and thickness of the leaky aquifer, respectively. Theoretically, leakage factor determines the radial distance from the well up to which the flow through the leaky layer is derived by the difference in hydraulic head and phreatic level. Indirectly, it determines the radius of influence of well. It has units of length.

2

Well Hydraulics

2.1 Introduction

During the pumping of aquifer, the flow behavior around the well can be analyzed by studying the governing equations derived for the situation. By incorporating the monitored pumping test data in the derived equations, the aquifer constants such as storage coefficient/specific yield transmissivity, and leakage factor are evaluated. This chapter includes the definitions related to well hydraulics, standard formulations, methods of evaluating formation constants, groundwater exploitation, and tube well design.

2.2 Definitions

Test hole – Drilling of a borehole through various geological formations to evaluate the site as potential well location is known as test hole. The number of test holes required depends upon the qualitative information collected from the already existing test hole(s) in the area. For proper sampling of groundwater, the diameter of the test hole should not be less than 3-inch. Normally, the test holes are drilled through the entire water-bearing formation.

Well sealing – When a material with very low permeability is placed into the well bore with the aim to prevent commingling of water between different aquifers and movement of water into an aquifer from the land surface, it is termed as well sealing.

Slotted/perforated casing – It is a lower-most portion of the well casing with openings for water entry.

Screen – A fabricated (locally) or manufactured well casing with precisely dimensioned and shaped openings.

Well inlet – The screened or perforated portion of the well casing through which water enters into the well.

Well intake zone – It is the portion of aquifer adjoining the well inlet that is modified by the well construction and development process. It includes the space between the well inlet and the undisturbed aquifer.

Pump column – It is the portion of well through which water from the aquifer is conveyed to the ground surface.

Water well – A hole excavated/drilled in the ground for bringing groundwater from the aquifer to the surface is known as water well.

Tube well – A pipe fixed for withdrawing water from aquifer, which passes through various geological formations consisting of pervious and impervious strata, is known as tube well.

Cavity well – A cavity well is a non-strainer well and draw water from the cavity formed below the confining layer (Figure 2.1a). A cavity well is installed in a geologic formation where a *thick clay layer* lies over an aquifer. While installing cavity well, the cutting shoe should be left about 60 cm above the bottom of the clay layer. Gradually, the movement of water through this 60 cm cylindrical bore converts its two ends in the bell-mouth shape, which reduces the entrance resistance (Figure 2.1b).

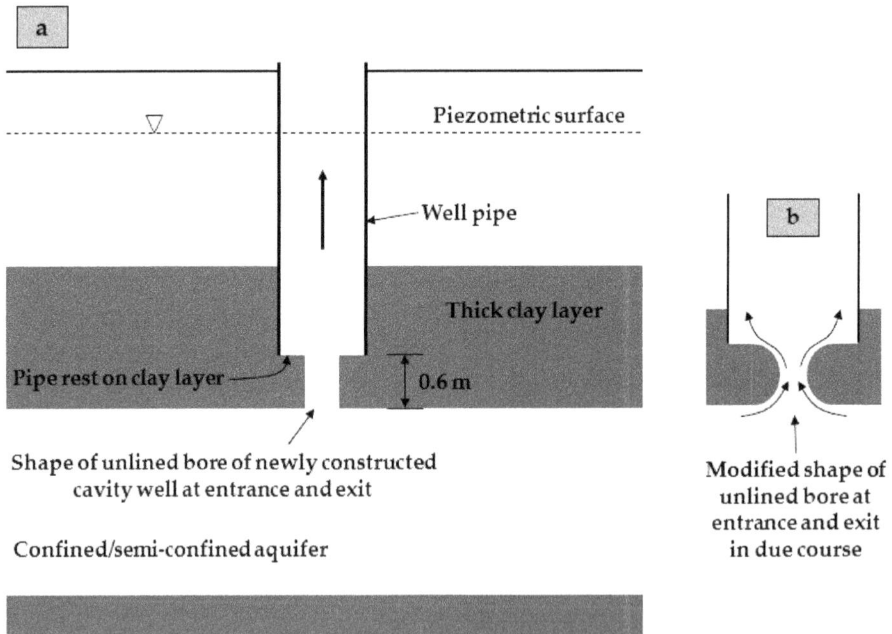

Figure 2.1: Schematic sketches of (a) cavity well, and (b) modification in the shape of unlined bore due to water movement.

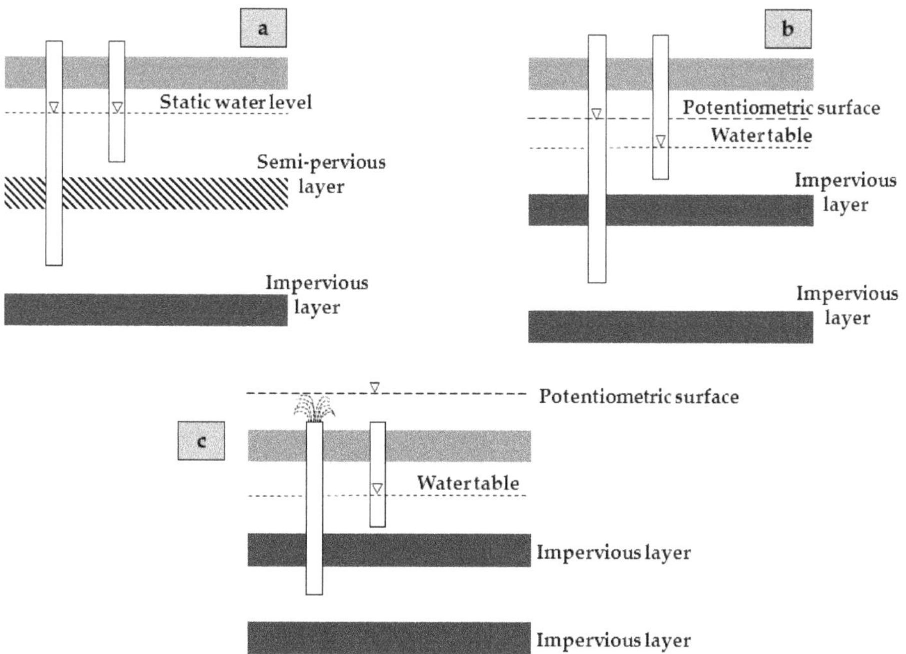

Figure 2.2: Schematic sketches of (a) phreatic-leaky aquifer system, (b) phreatic-confined aquifer system, and (c) a flowing well.

Static water level – Static water level indicates the level of water in the well before pumping starts. In a phreatic-leaky aquifer system, water table levels with potentiometric surface (Figure 2.2a). For the wells installed in the completely confined aquifers, static water level may differ appreciably from the water table (see Figure 2.2b for comparison). In case of *flowing wells*, the static water level lies above the ground surface (Figure 2.2c). The pressure of water at the static water level is atmospheric.

Piezometric level – Piezometric level is the height (h) at which water will stand in a piezometer, which is installed in a confined/semi-confined aquifer. The rise of water in the pipe is due to the pressure (p) of water at the open end of pipe and can be evaluated using the formula $h = \frac{p}{w}$. Where, w is the unit weight of water.

Pumping water level – The level of water in the well during pumping of an aquifer is pumping water level.

Drawdown – Drawdown in the well at any instant is the difference between static water level/potentiometric surface and pumping water level (Figures 2.3-2.4). The maximum possible drawdown in a well is decided by the pumping water level, which cannot be lower than the top of well screen.

During the pumping of an unconfined aquifer, an inverted cone separates the saturated part of the aquifer from the drained portion. This inverted cone has static

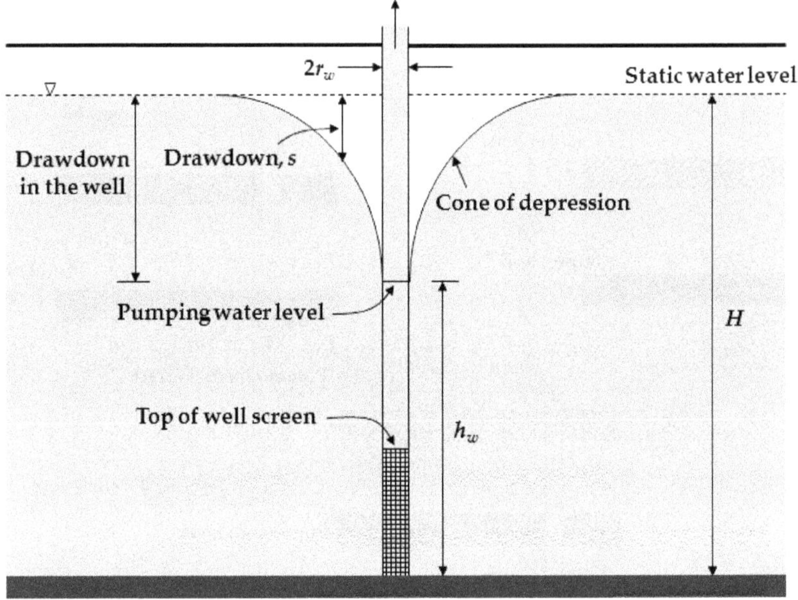

Figure 2.3: Schematic sketch showing drawdown around the water table well.

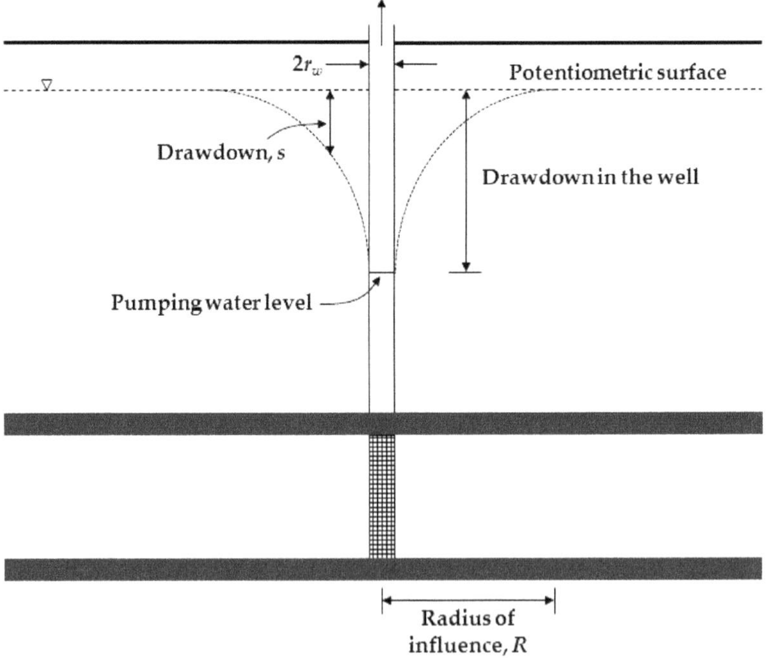

Figure 2.4: Schematic sketch showing drawdown distribution around the artesian well.

water level as *base* and pumping water level as *apex* and is also known as *cone of depression*. However, in case of pumping of the confined/semi-confined aquifers, the inverted cone represents the piezometric head distribution with the radial distance and can be considered *imaginary* (Figure 2.4). The base area of the inverted cone is called the *area of influence*. The boundary covering the area of influence is known as *circle of influence* and whose radius is termed as *radius of influence*.

Well yield – Well yield represents the volume of water pumped per unit time from it.

Specific capacity – Specific capacity of a well at any instant is evaluated by dividing well yield by drawdown at the well face.

Well test – It is performed to determine the well yield and change in drawdown with respect to time. The observed data is used to evaluate the aquifer constants.

Well efficiency – The ratio of theoretical drawdown to measured drawdown is known as well efficiency.

Filter points – Filter points is a term used to express shallow tube wells in deltaic region.

Perennial yield – It is the rate at which water may be withdrawn from a groundwater basin without producing an undesirable effect.

2.3 General flow equations

2.3.1 *Rectangular coordinates* – The general flow equation in three-dimensions can be written as

$$K_x \frac{\partial^2 h}{\partial x^2} + K_y \frac{\partial^2 h}{\partial y^2} + K_z \frac{\partial^2 h}{\partial z^2} = S \frac{\partial h}{\partial t} \tag{2.1}$$

In Eq. (2.1), K_x, K_y, K_z are the hydraulic conductivities in x, y, and z directions, respectively; h is the hydraulic head; S is the coefficient of storage; and t is the time. Under steady state conditions, the term $\frac{\partial h}{\partial t}$ is equal to zero and the Eq. (2.1) changes to

$$K_x \frac{\partial^2 h}{\partial x^2} + K_y \frac{\partial^2 h}{\partial y^2} + K_z \frac{\partial^2 h}{\partial z^2} = 0 \tag{2.2}$$

For saturated homogeneous and isotropic porous formations, $K_x = K_y = K_z = K$. Therefore, the Eq. (2.2) modifies to

$$\frac{\partial^2 h}{\partial x^2} + \frac{\partial^2 h}{\partial y^2} + \frac{\partial^2 h}{\partial z^2} = 0 \tag{2.3}$$

The Eq. (2.3) is designated as the Laplace equation (Freeze and Witherspoon, 1966).

2.3.2 *Radial coordinates* – During groundwater pumping through wells, flow is axisymmetric. It is convenient to express the flow equation in radial coordinates. The equation for radial flow in a homogeneous and isotropic aquifer can be written as

$$\frac{\partial^2 h}{\partial r^2} + \frac{1}{r} \cdot \frac{\partial h}{\partial r} = \frac{S}{T} \cdot \frac{\partial h}{\partial t} \tag{2.4}$$

In Eq. (2.4), r is the radial coordinate from the well center; and T is the transmissivity of the aquifer. Steady state conditions are reached when aquifer is pumped for considerable time. In that situation, Eq. (2.4) modifies to

$$\frac{\partial^2 h}{\partial r^2} + \frac{1}{r} \cdot \frac{\partial h}{\partial r} = 0 \tag{2.5}$$

2.3.3 *Steady unidirectional flow in a confined aquifer* – The Figure 2.5 shows a cross-section through a confined aquifer with hydraulic conductivity, K. Say, the piezometric heads at locations A (i.e. $x = 0$) and B (i.e. $x = L$) monitored through piezometers are h_A and h_B, respectively. Therefore, the general equation for evaluating hydraulic head at any point within the domain can be obtained as

$$h_x = h_A + \left(h_B - h_A\right) \cdot \frac{x}{L} \tag{2.6}$$

Equation (2.6) describes that the head decreases linearly (Figure 2.5).

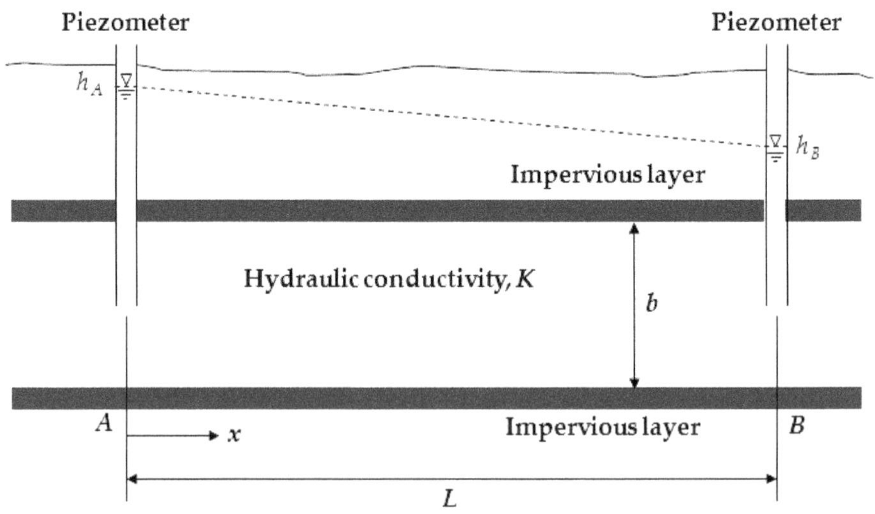

Figure 2.5: Schematic sketch showing the groundwater flow through a confined aquifer.

2.4 Mathematical formulations for evaluating steady flow to wells

2.4.1 *Steady-state radial flow to a well in confined aquifer* – All the variables required to derive the steady state flow equation towards a fully penetrating well installed in a confined aquifer are depicted in Figure 2.6. Since, the flow towards well is two-dimensional, the Dupuit assumptions are valid. Consider a cylindrical volume element around the well. The hydraulic gradient at any radial distance (r) can be evaluated by finding out the slope of tangent to drawdown curve (Figure 2.6). The mathematical equation for evaluating the flow crossing the cylindrical surface (i.e. at radial distance r) can be written as

$$Q_w = -K \cdot \frac{dh}{dr} \cdot 2\pi rb \tag{2.7}$$

Integrating the Eq. (2.7) within the limits as $r = r_w$, $h = h_w$ and $r = R$, $h = H$ after separating the hydraulic head and radial distance terms (neglecting negative sign), we get

$$Q_w = 2\pi Kb \frac{(H - h_w)}{\log_e \frac{R}{r_w}} \tag{2.8}$$

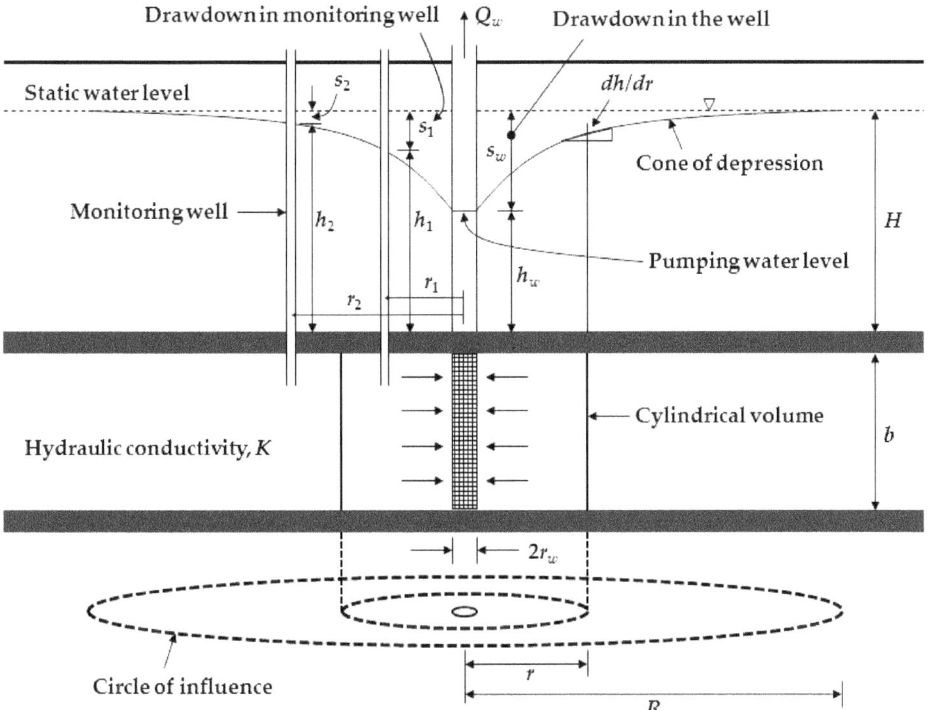

Figure 2.6: Schematic sketch of a pumping well installed in a confined aquifer.

As we know that $H - h_w = s_w$, the Eq. (2.8) modifies to

$$Q_w = 2\pi Kb \frac{s_w}{\log_e \frac{R}{r_w}} \tag{2.9}$$

Equation (2.9) is also known as *equilibrium* or *Thiem* equation. Consider that h_1 and h_2 are the hydraulic heads observed in two monitoring wells installed at radial distances of r_1 and r_2, respectively from the well center. Modified form of Eq. (2.8) for these changes will be

$$Q_w = 2\pi Kb \frac{(h_2 - h_1)}{\log_e \frac{r_2}{r_1}} \tag{2.10}$$

In actual field conditions, drawdown instead of head is monitored. From Figure 2.6, it is clear that $h_1 = H - s_1$ and $h_2 = H - s_2$. Incorporating these values in Eq. (2.10), it changes to

$$Q_w = 2\pi Kb \frac{(s_1 - s_2)}{\log_e \frac{r_2}{r_1}} \tag{2.11}$$

Equation (2.11) can be applied accurately only when the pumping well has approached the steady-state conditions. In other words, it is a state when there is no appreciable change in drawdown with time.

2.4.2 *Steady-state radial flow to a well in unconfined aquifer* – All the variables required to derive the steady state flow equation towards a well installed in an unconfined aquifer are depicted in Figure 2.7.

The well is penetrating the unconfined aquifer up to the confining layer of lower aquifer. The equation for groundwater flow at any radial distance (r) can be written using Darcy's law as

$$Q_w = -K \cdot \frac{dh}{dr} \cdot 2\pi rh \tag{2.12}$$

Integrating the Eq. (2.12) within the limits as $r = r_w$, $h = h_w$ and $r = R$, $h = H$ after separating the hydraulic head and radial distance terms (neglecting negative sign), we get

$$Q_w = \pi K \frac{\left(H^2 - h_w^2\right)}{\log_e \frac{R}{r_w}} \tag{2.13}$$

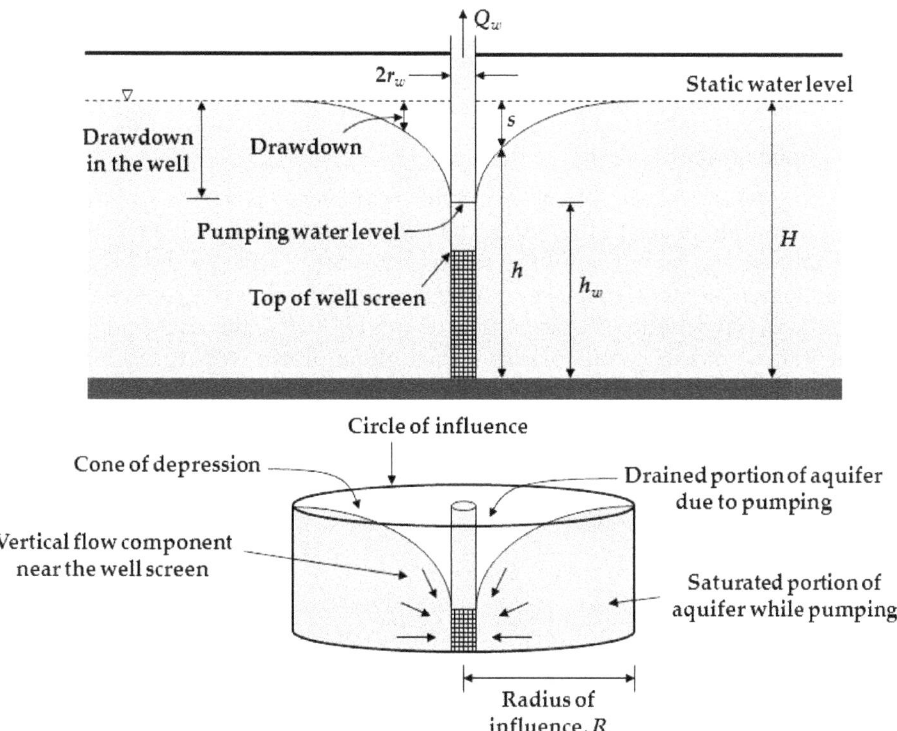

Figure 2.7: Schematic sketch of a pumping well installed in an unconfined aquifer.

Writing the generalized equation for the two monitoring wells installed at radial distances of r_1 and r_2, which are observing hydraulic heads h_1 and h_2, respectively as

$$Q_w = \pi K \frac{\left(h_2^2 - h_1^2\right)}{\log_e \frac{r_2}{r_1}} \tag{2.14}$$

Let s_1 and s_2 be the steady-state drawdowns at two monitoring wells installed at radial distances of r_1 and r_2, respectively. By incorporating $h_1 = H - s_1$ and $h_2 = H - s_2$ in Eq. (2.14) and thereafter replacing the terms $s_1 - \frac{s_1^2}{2H}$ and $s_2 - \frac{s_2^2}{2H}$ with s_1' and s_2' based on the condition that drawdowns are small (in comparison to saturated thickness of aquifer, H), it changes to

$$Q_w = 2\pi KH \frac{\left(s_1' - s_2'\right)}{\log_e \frac{r_2}{r_1}} \tag{2.15}$$

In Eq. (2.15) s'_1 and s'_2 are the *corrected drawdowns* and *KH* is the average aquifer transmissivity. Comparison of the Eq. (2.15) with Eq. (2.11) shows that these are identical. Therefore, conclusion can be drawn that for small drawdowns (in comparison to saturated thickness of aquifer), a phreatic aquifer may be treated as a confined aquifer.

2.4.3 *Steady-state radial flow to a well in unconfined aquifer with uniform recharge* – All the variables required to derive the steady state flow equation towards a well installed in an unconfined aquifer with uniform recharge are depicted in Figure 2.8. Consider a volume element at radial distance (r) from the well. The flow (Q_a) enters the volume element at radial distance ($r + dr$) and leaves as $Q_a + dQ_a$ at radial distance (r). The change in flow rate dQ_a can be attributed to the uniform recharge rate v from the above. The equation for contribution from the uniform recharge can be written as

$$dQ_a = -(2\pi r\, dr)v \tag{2.16}$$

Integrating the Eq. (2.16), we get

$$Q_a = -\pi r^2 v + C_1 \tag{2.17}$$

In Eq. (2.17), C_1 is an arbitrary constant. Incorporating the condition that when $r \to 0$, $Q_a \to Q_w$ the value of C_1 comes equal to Q_w. After incorporating the value of C_1 in Eq. (2.17), it changes to

$$Q_a = Q_w - \pi r^2 v \quad or \quad Q_a = Q_w - Q_v \tag{2.18}$$

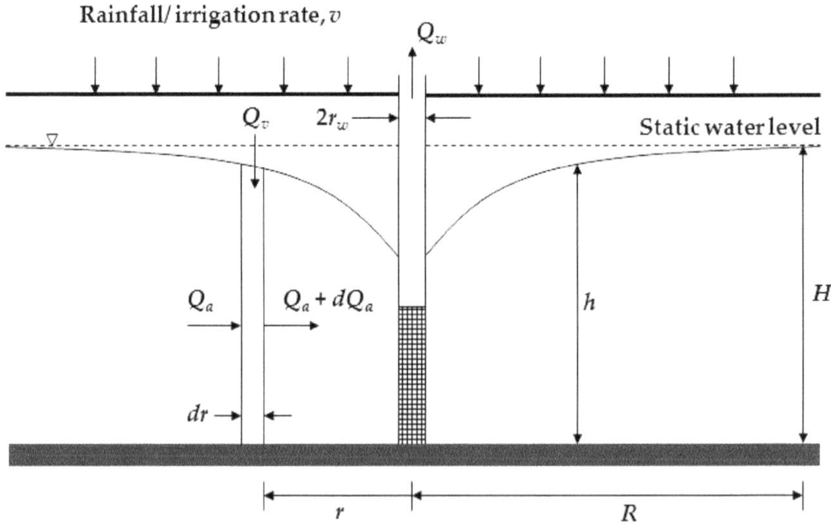

Figure 2.8: Schematic sketch showing the steady groundwater flow to a well installed in a uniformly recharged unconfined aquifer.

Substituting Eq. (2.12) as the aquifer contribution (i.e. Q_a) to total well discharge (i.e. Q_w) in Eq. (2.18), we get

$$K \cdot \frac{dh}{dr} \cdot 2\pi rh = Q_w - \pi r^2 v \tag{2.19}$$

On integrating the Eq. (2.19) after separating the hydraulic head and radial distance terms, it changes to

$$h^2 = \frac{Q_w}{\pi K} \log_e r - \frac{v}{2K} r^2 \tag{2.20}$$

From Figure 2.8, it is clear that when $r \rightarrow R,\ h \rightarrow H$. Incorporating this condition in Eq. (2.20), it modifies to

$$H^2 = \frac{Q_w}{\pi K} \log_e R - \frac{v}{2K} R^2 \tag{2.21}$$

On subtracting Eq. (2.20) from Eq. (2.21), the generalized equation for drawdown curve can be written as

$$H^2 - h^2 = \frac{Q_w}{\pi K} \log_e \frac{R}{r} - \frac{v}{2K}\left(r^2 - R^2\right) \tag{2.22}$$

2.4.4 *Radial flow in a confined aquifer under unsteady state conditions* – During the pumping of a radially extensive aquifer, the influence of discharge continue to extend outward with time. This leads to the development of unsteady state flow conditions in the aquifer. Consider a well is fully penetrating an aquifer, which is completely confined, homogeneous, isotropic, of uniform thickness, and infinite in aerial extent. The applicable differential equation for unsteady radial flow towards a well in a confined aquifer in plane polar coordinates can be written as

$$\frac{\partial^2 s}{\partial r^2} + \frac{1}{r} \cdot \frac{\partial s}{\partial r} = \frac{S}{T} \cdot \frac{\partial s}{\partial t} \tag{2.23}$$

In Eq. (2.23), s is the drawdown; r is the radial distance from the well center; S is the storage coefficient; T is aquifer transmissivity; and t is time. *Theis* has evaluated the solution of Eq. (2.23) by incorporating the initial and boundary conditions $s = s_0\ at\ t = 0$ and $s \rightarrow s_0\ as\ r \rightarrow \infty$. The solution to Eq. (2.23) can be written as

$$s = \frac{Q_w}{4\pi T} \int_u^\infty \frac{e^{-u}}{u} du \quad or \quad s = \frac{Q_w}{4\pi T} W(u) \quad with \quad u = \frac{r^2 S}{4Tt} \tag{2.24}$$

Equation (2.24) is also known as the *nonequilibrium*. The term $W(u)$ is known as *well function*. The integral term in Eq. (2.24) is known as *exponential integral* and is a function of u. Equation (2.24) in expanded form can be written as

$$s = \frac{Q_w}{4\pi T}\left[-0.5772 - \ln u + u - \frac{u^2}{2\times 2!} + \frac{u^3}{3\times 3!} - \frac{u^4}{4\times 4!} + \cdots\right] \tag{2.25}$$

2.5 Methods for evaluating formation constants

2.5.1 *Theis method* – It is based upon the graphic method of superposition for evaluating formation constants. Re-writing the Eq. (2.24) by inserting parenthesis as

$$s = \left(\frac{Q_w}{4\pi T}\right)W(u) \tag{2.26}$$

Rearrange the variables of u (Eq. 2.24) in the following form

$$\frac{r^2}{t} = \left(\frac{4T}{S}\right)u \tag{2.27}$$

The terms in parenthesis of Eq.s (2.26) and (2.27) are constant. There is a particular relation between the function $W(u)$ and u. Therefore, a similar relation can be think of between the terms s and $\frac{r^2}{t}$. The well function $W(u)$ is plotted against u on a logarithmic paper with u on abscissa and $W(u)$ on ordinate. The plot is also known as *type curve* (Figure 2.9).

Keeping the same scale, $\frac{r^2}{t}$ (on abscissa) is plotted against drawdown, s (on ordinate). The observed time-drawdown plot is superimposed on the type curve, keeping the abscissa and ordinate of one plot parallel to abscissa and ordinate of the other. Both the plots are adjusted in a manner so that most of the plotted points of the observed data fall on a segment of the type curve (Figure 2.10). Consider any point on the matched curves and draw horizontal and vertical lines crossing abscissas and ordinates of both the graphs. The values so obtained for $W(u)$, u, s, and $\frac{r^2}{t}$ are incorporated in Eq.s (2.26) and (2.27) to find out S and T.

2.5.2 *Cooper-Jacob method* – The value of u at smaller radial distance from the well after a prolonged pumping period is small. As a result, the values of the series terms in Eq. (2.25) become negligible after the first two terms. With the mentioned condition, Eq. (2.25) modifies to

$$s = \frac{Q_w}{4\pi T}\times\left[-0.5772 - \log_e \frac{r^2 S}{4Tt}\right] \tag{2.28}$$

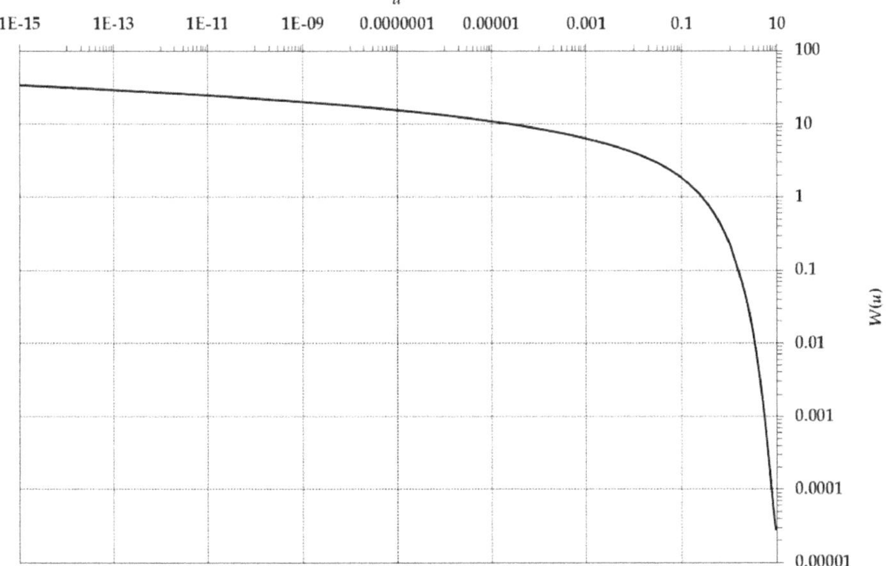

Figure 2.9: A plot of *W*(*u*) against *u* on double-logarithmic paper (type curve).

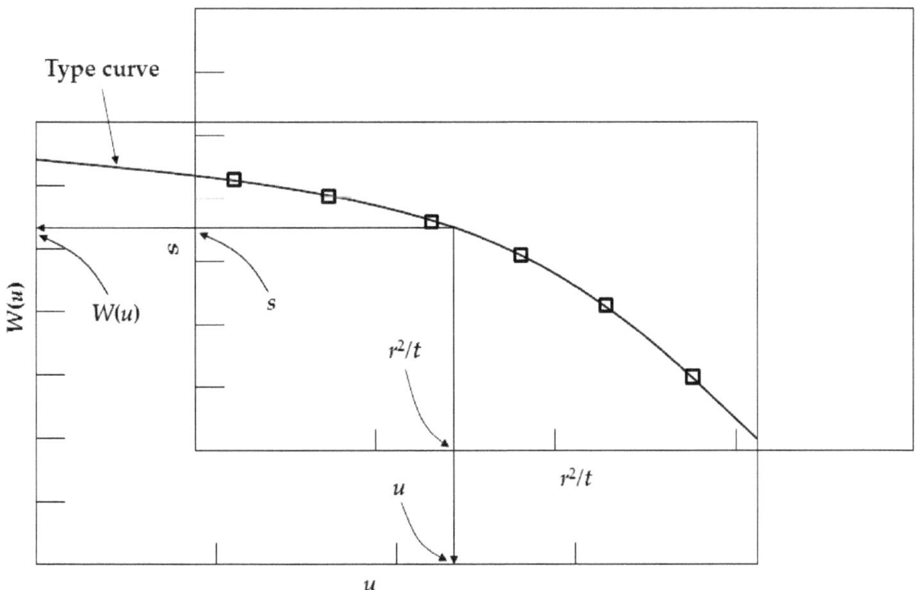

Figure 2.10: Theis method of superposition for evaluating formation constants.

After combining the terms in parenthesis and converting the natural logarithm to base-10 logarithm, Eq. (2.28) modifies to

$$s = \left(\frac{2.303 Q_w}{4\pi T}\right) \times \log_{10}\left(\frac{2.25T}{r^2 S}\right) t \tag{2.29}$$

Therefore, the plot between drawdown (s) and logarithm of time (t) will be a straight line (Figure 2.11). If not, it may be fitted. The straight line is projected back to time-axis. The drawdown (s) will be equal to zero at the meeting point and may be symbolized as t_0. Incorporating the mentioned condition (i.e. $t = t_0$ at $s = 0$) in Eq. (2.29) leads it to

$$0 = \left(\frac{2.303 Q_w}{4\pi T}\right) \times \log_{10}\left(\frac{2.25T}{r^2 S}\right) t_0 \tag{2.30}$$

Since, the first term on right-hand side of Eq. (2.30) is not equal to zero; it can be re-written as

$$S = \frac{2.25 T t_0}{r^2} \tag{2.31}$$

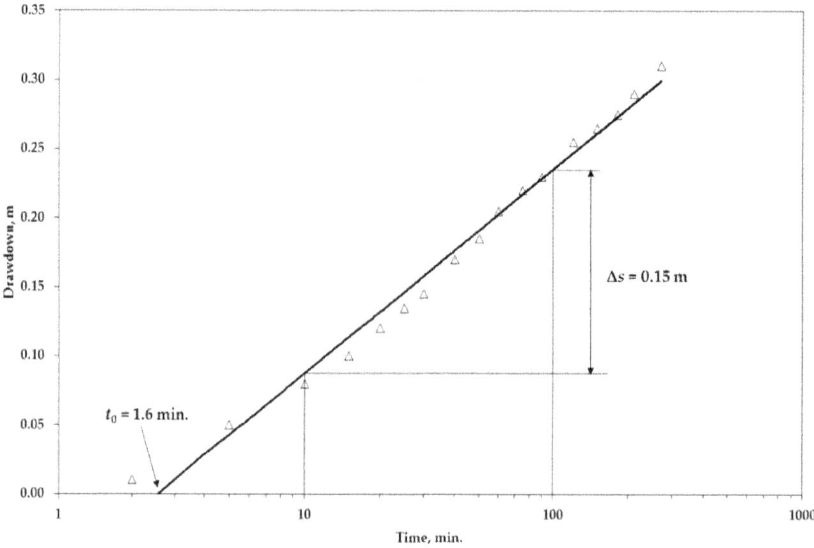

Figure 2.11: Cooper-Jacob method for evaluating aquifer constants.

Incorporation of Eq. (2.31) in Eq. (2.29) leads to

$$s = \left(\frac{2.303\,Q_w}{4\pi T}\right) \times \log_{10}\left(\frac{t}{t_0}\right) \tag{2.32}$$

The Eq. (2.32) can easily be solved by noting change in drawdown (Δs) for one-log cycle (starting from t_0) on the time-drawdown curve (Figure 2.11). Therefore, by incorporating $\frac{t}{t_0} = 10$ and $s = \Delta s$ in Eq. (2.32), the formulation for evaluating transmissivity (T) can be written as

$$T = \frac{2.303\,Q_w}{4\pi\,\Delta s} \tag{2.33}$$

Since, the Eq. (2.33) is independent of storage coefficient (S), the value of transmissivity (T) is evaluated first. This method is proposed by Cooper and Jacob (1946). To avoid large errors in the evaluated constants, the value of u must be less than 0.01.

2.5.3 *Chow method* – Similar to Cooper and Jacob method of solution, the first step is to plot the observed drawdown data against time on semi-logarithmic paper. A point is chosen on the curve and corresponding drawdown value (s) is noted by drawing horizontal line (Figure 2.12).

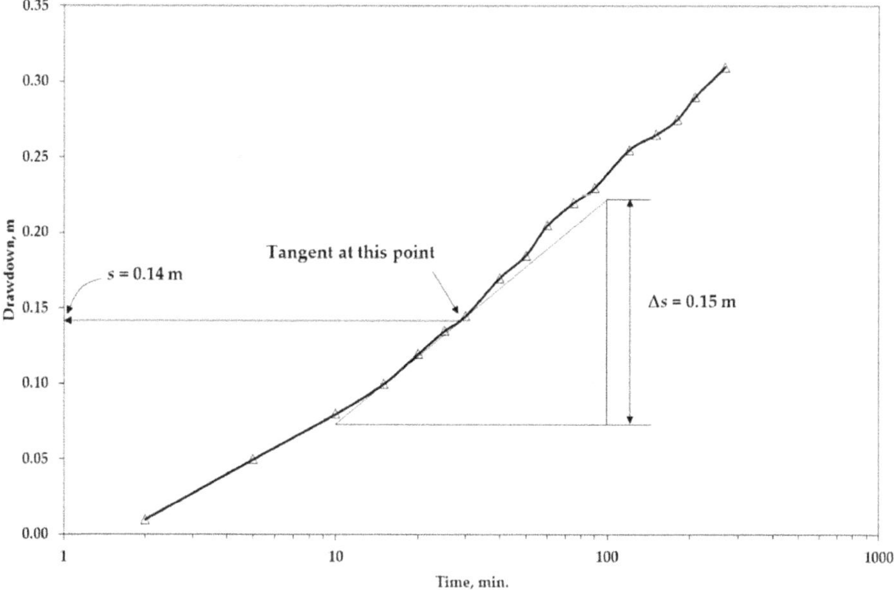

Figure 2.12: Chow's method for evaluating formation constants.

Moreover, the tangent is drawn at the chosen point. Difference in drawdown Δs for the one-log cycle of time is evaluated. From the determined values, the function $F(u)$ is evaluated as

$$F(u) = \frac{s}{\Delta s} \tag{2.34}$$

For the evaluated value of $F(u)$, corresponding values of u and $W(u)$ are determined from the Figure 2.13. The determined values are incorporated in the Eq.s (2.26) and (2.27) for evaluating transmissivity (T) and storage coefficient (S).

2.5.4 *Recovery test method* - After stopping the pump, the rise of water level in the pumped aquifer is referred to as the recovery of groundwater. The measurement of drawdown during the recovery period is known as *residual drawdown* (Figure 2.14). Rate of recharge during the recovery period is considered constant and equal to the mean pumping rate Q_w. According to Theis (1935), the residual drawdown (s') can be evaluated mathematically as

$$s' = \frac{Q_w}{4\pi T}\left[W(u) - W(u')\right] \quad with \quad u = \frac{r^2 S}{4Tt} \quad and \quad u' = \frac{r^2 S}{4Tt'} \tag{2.35}$$

In Eq. (2.35), t is the pumping period and t' is the time after the shutting down of the pump. Solving the Eq. (2.35) after incorporating the expansion terms of $W(u)$

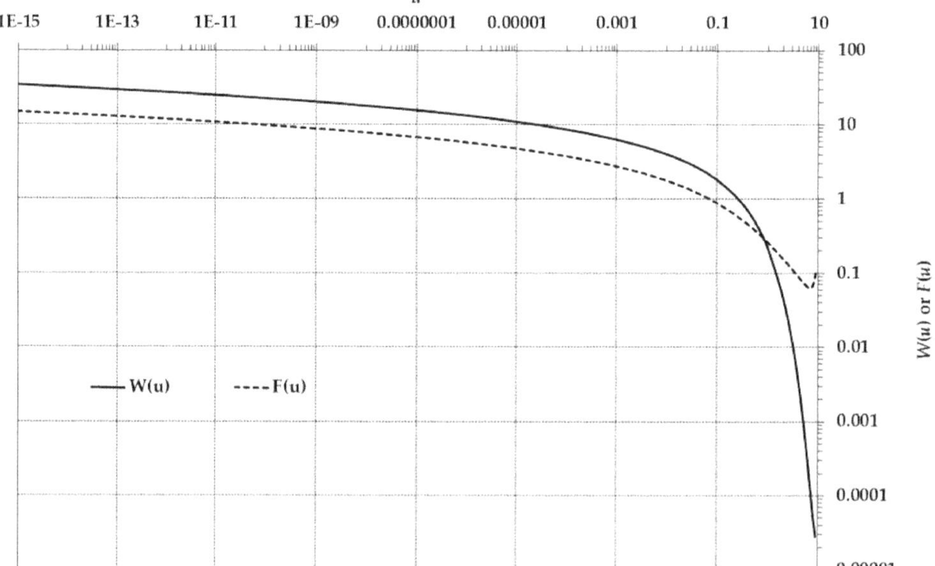

Figure 2.13: Relation between *F*(*u*), *W*(*u*), and *u* (Chow, 1952).

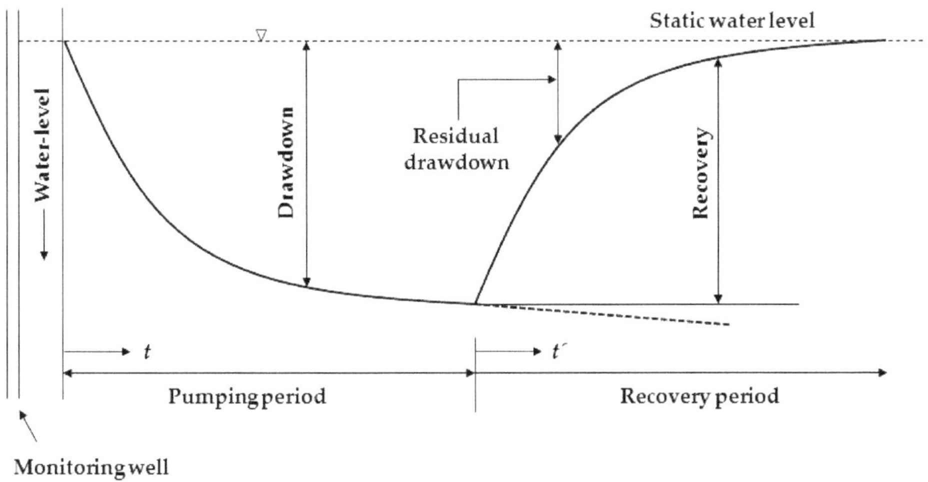

Figure 2.14: Drawdown and recovery curve.

and $W(u')$ (subject to the condition that the drawdown is monitored near the well after prolonged pumping period), it changes to

$$s' = \frac{2.303Q_w}{4\pi T}\log_{10}\frac{t}{t'} \tag{2.36}$$

The plot of residual drawdown (s') against the $\log_{10}\frac{t}{t'}$ forms a straight line (Figure 2.15). Therefore, for one-log cycle, the change in drawdown will be equal to slope of the line. Hence, the Eq. (2.36) changes to

$$\Delta s' = \frac{2.303Q_w}{4\pi T}\log_{10}10 \quad \rightarrow \quad T = \frac{2.303Q_w}{4\pi \Delta s'} \tag{2.37}$$

From the recovery period data, accurate value of transmissivity can be obtained but storage coefficient value appreciably varies from the correct result.

2.6 Unsteady radial flow towards well in leaky aquifer

During the pumping of a leaky aquifer, water is withdrawn both from the aquifer and from the overlying aquifer through the aquitard (or semi-pervious layer). The quantity of water that will pass through the aquitard at any point will be proportional to the difference between hydraulic head values above and bottom of the aquitard layer (Figure 2.16). The drawdown of the piezometric surface after the start of pumping from a well installed in a leaky aquifer can be written as

$$s = \frac{Q_w}{4\pi T}W\left(u, \frac{r}{B}\right) \quad \textit{with} \quad u = \frac{r^2 S}{4Tt} \quad \textit{and} \quad \frac{r}{B} = \frac{r}{\sqrt{K\,bc}} \quad \textit{where} \quad c = \frac{b'}{K'} \tag{2.38}$$

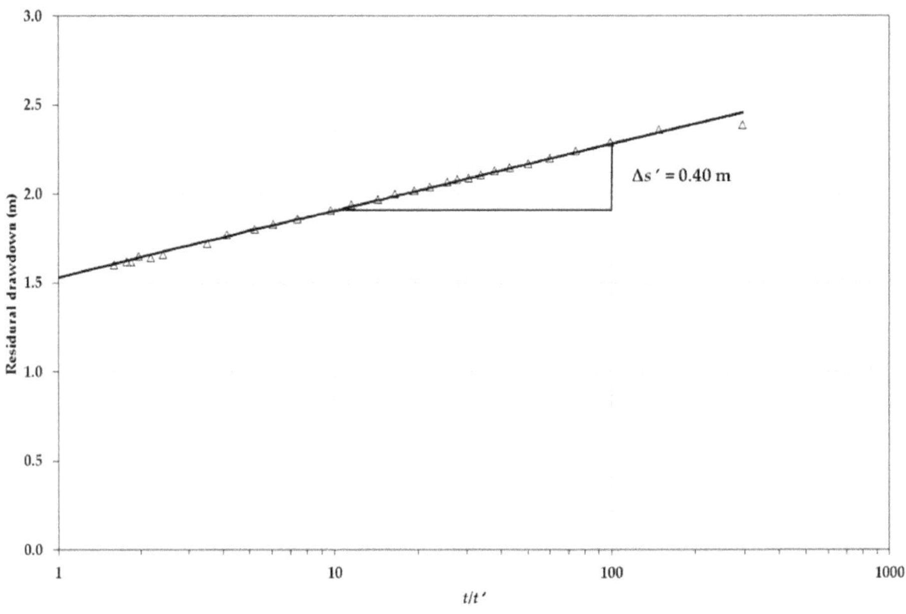

Figure 2.15: Recovery test method for evaluating transmissivity of aquifer.

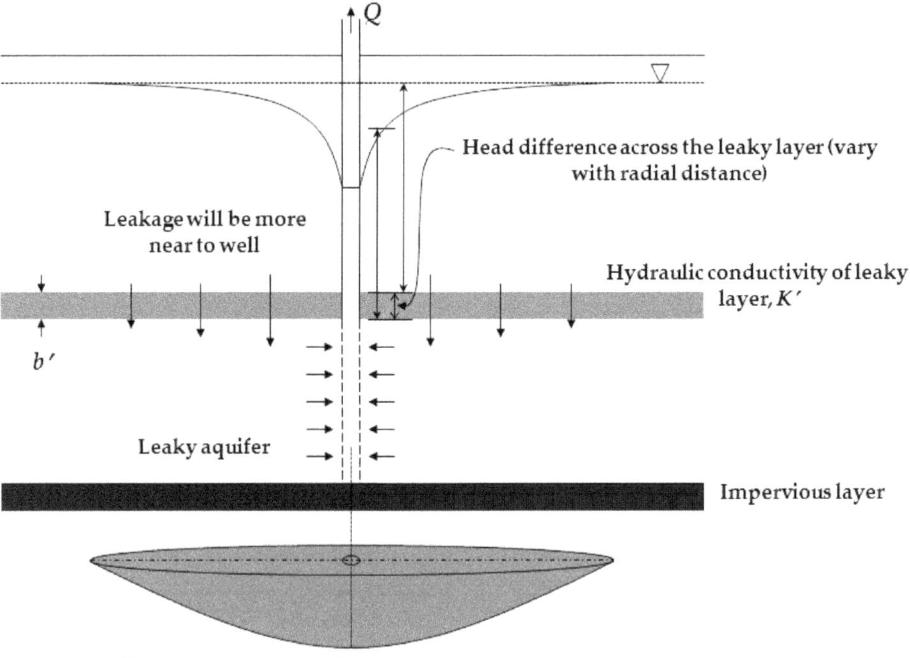

Figure 2.16: Schematic sketch of a pumping well in a leaky aquifer.

In Eq. (2.38), K' and b' are the hydraulic conductivity and thickness of the leaky layer, respectively. For a completely confined aquifer

$$K' = 0 \quad \Rightarrow \quad c = \infty \quad \Rightarrow \quad \frac{r}{B} = 0 \tag{2.39}$$

On incorporating $\frac{r}{B} = 0$ in Eq. (2.38), it changes to *Theis* equation (Eq. 2.24).

2.7 Partially penetrating wells

The wells, which do not penetrate the full aquifer thickness are called *partially penetrating* wells (Figure 2.17a). The wells, which fully penetrate the aquifer but only partially screen it, are known as partially-screening/partially-tapping wells (Figure 2.17b). The groundwater flow pattern towards partially penetrating wells is shown schematically and is compared with the radial flow in case of fully penetrating well (Figure 2.18). The longer flow lines in case of partially penetrating wells cause greater resistance to flow and more head loss. Consider two wells (one fully and other partially penetrating) are installed in an aquifer under similar conditions. If both the wells are discharging with same rate, then the drawdown at the well face for partially penetrating well will be more in comparison to fully penetrating well (Figure 2.19). Conversely, if well face drawdown is same for the two wells then surely, the partially penetrating well will discharge with lesser rate than fully penetrating well.

Consider a well is partially penetrating an infinite leaky artesian aquifer (Figure 2.20). The origin of radial and vertical coordinates is also shown in the figure. The direction of arrows is showing the positive direction. The flow towards the well may be defined by the following boundary value problem

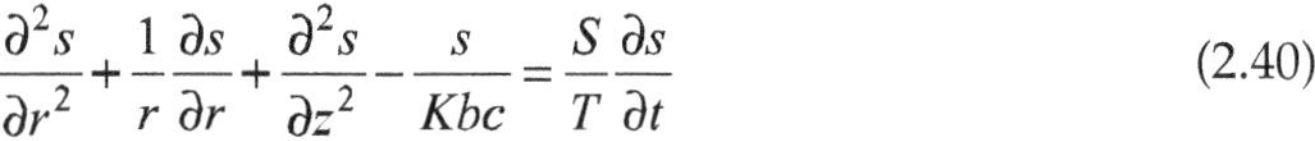

$$\frac{\partial^2 s}{\partial r^2} + \frac{1}{r}\frac{\partial s}{\partial r} + \frac{\partial^2 s}{\partial z^2} - \frac{s}{Kbc} = \frac{S}{T}\frac{\partial s}{\partial t} \tag{2.40}$$

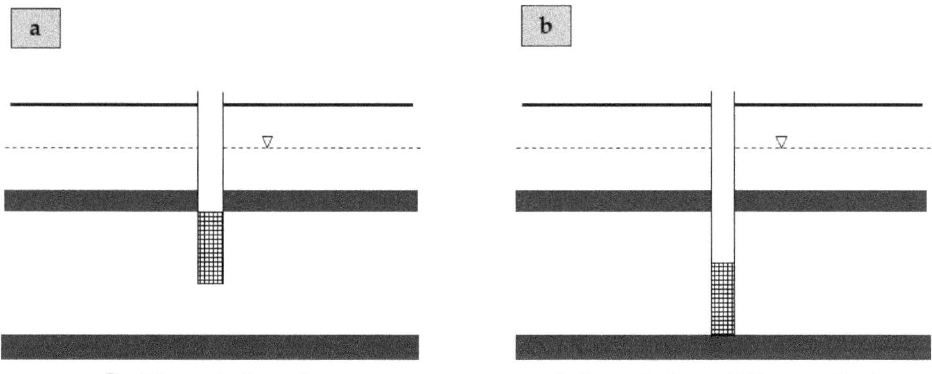

Figure 2.17: Schematic sketches showing the difference between partially penetrating and fully penetrating partially screened well.

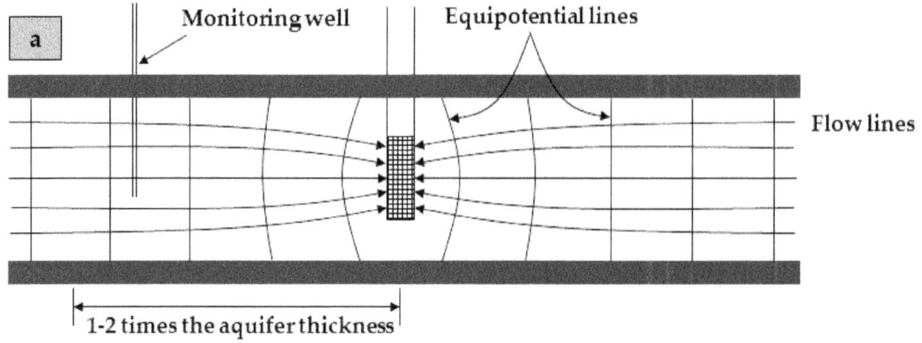

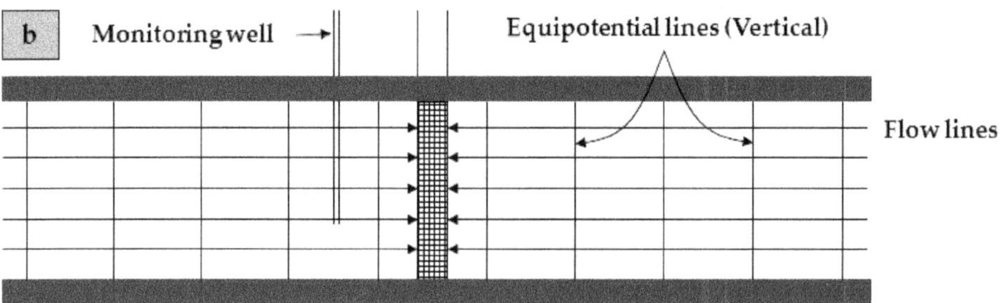

Figure 2.18: Schematic sketches showing the effect of (a) partially penetrating well, and (b) fully penetrating well on the placement of monitoring well.

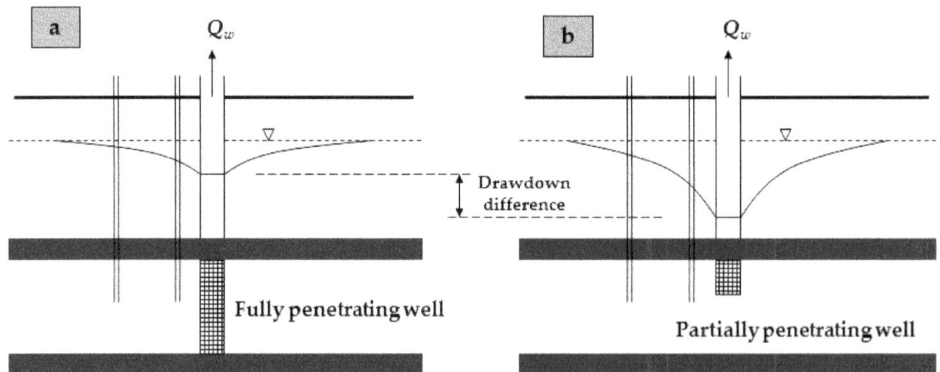

Figure 2.19: Schematic sketches showing the drawdown difference between partially and fully penetrating wells while pumping the aquifer with same rate.

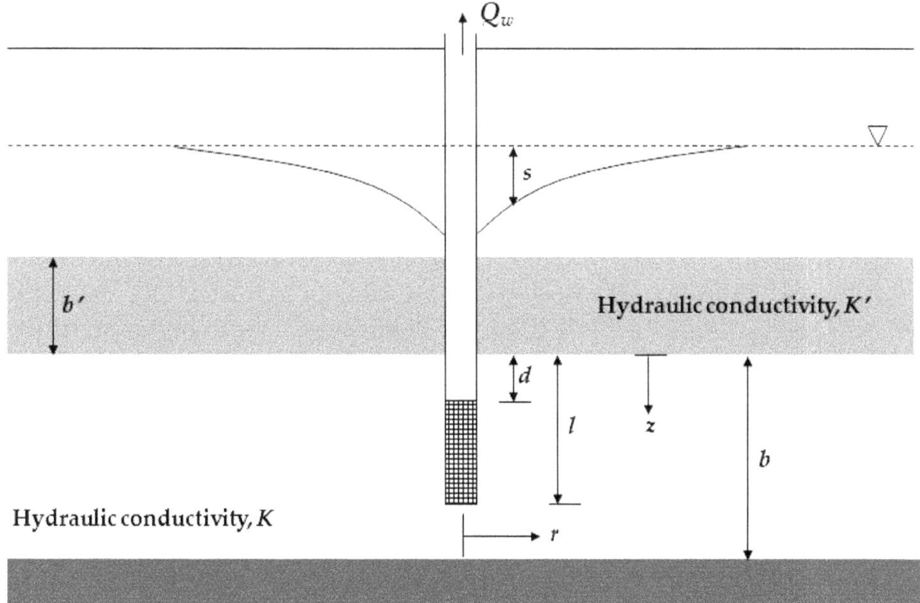

Figure 2.20: A partially penetrating well in a leaky aquifer.

With initial condition $s(r, z, t = 0) = 0;$ (2.41)

And with boundary conditions $s(r \to \infty, z, t) = 0;$ (2.42)

$$\frac{\partial s(r, z = 0, t)}{\partial z} = \frac{\partial s(r, z = b, t)}{\partial z} = 0; \tag{2.43}$$

$$\lim_{r \to 0} r \cdot \frac{\partial s}{\partial r} = 0 \qquad 0 < z < d \tag{2.44a}$$

$$\lim_{r \to 0} r \cdot \frac{\partial s}{\partial r} = -\frac{Q_w}{2\pi K(l - d)} \qquad d < z < l \tag{2.44b}$$

$$\lim_{r \to 0} r \cdot \frac{\partial s}{\partial r} = 0 \qquad l < z < b \tag{2.44c}$$

Applying Laplace and finite Fourier cosine transformations, Hantush (1964) has given solution to the boundary value problem (Eq.s 2.40-2.44) as

$$s = \frac{Q_w}{4\pi Kb}\left[W\left(u, \frac{r}{B}\right) + \left\{\frac{2b}{\pi(l - d)}\right\}\sum_{n=1}^{\infty} R_n \cdot W\left(u, \sqrt{\left(\frac{r}{B}\right)^2 + \left(\frac{n\pi r}{b}\right)^2}\right)\right] \tag{2.45}$$

Where, $R_n = \left(\frac{1}{n}\right) \cdot \left[\text{Sin}\frac{n\pi l}{b} - \text{Sin}\frac{n\pi d}{b}\right] \cdot \text{Cos}\frac{n\pi z}{b}$

All the variables described in Eq. (2.45) are depicted through Figure 2.20 and are defined in the previous sections.

2.8 Characteristic well losses

The drawdown at a well is the sum of logarithmic drawdown curve at the well face and the well loss, which is caused by the flow through well screen and flow inside the well pipe to pump intake. It is because of this reason the pumping water level inside the well is little lower than the level where drawdown curve meets at the well face (Figure 2.21). The basic reason of well loss is turbulent flow and therefore, it will be proportional to some n^{th} power of well discharge as Q_w^n. Where, n is a constant having value greater than one.

While pumping a confined aquifer under steady state conditions, the total drawdown s_w (including the well loss) can be written from Eq. (2.9) as

$$s_w = \frac{Q_w}{2\pi Kb} \log_e \frac{R}{r_w} + CQ_w^n \tag{2.46}$$

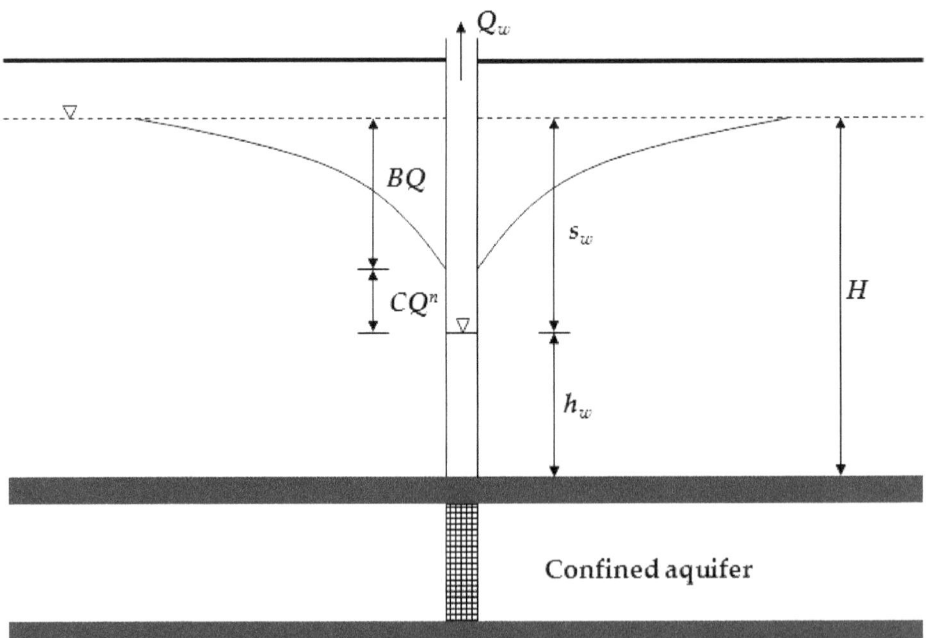

Figure 2.21: Schematic sketch showing the formation and screen loss.

In Eq. (2.46), C is a constant whose value depends upon the radius, construction, and condition of the well. Considering $\left(\frac{\log_e \frac{R}{r_w}}{2\pi Kb}\right) = B$, Eq. (2.46) changes to

$$s_w = BQ_w + CQ_w^n \tag{2.47}$$

In Eq. (2.47), the term BQ_w is regarded as formation loss, whereas the term CQ_w^n is because of well loss (Figure 2.21). With the deterioration or clogging of screen, increase in well losses has been noticed.

2.8.1 *Evaluation of well loss* – Well loss can be evaluated by step-drawdown pumping test. Initially, pumping is started at low-rate and continued until the drawdown stabilizes. Following the similar pattern, drawdown data is procured by increasing the discharge rate step-wise. For $n = 2$, Eq. (2.47) changes to

$$\frac{s_w}{Q_w} = B + CQ_w \tag{2.48}$$

A straight line is fitted on the points plotted between $\frac{s_w}{Q_w}$ vs. CQ_w. The slope of the line will give the well loss coefficient C and the formation loss coefficient B by the intercept $Q_w = 0$. Later on, this graphical analysis is modified by the Rorabaugh (1953) for the conditions when the value of n significantly deviates from 2.

2.9 Groundwater exploitation in hard rock areas

The hard rock areas generally have poor permeability and therefore, bore wells and tube wells are usually unsuitable in such formations. Open wells are more suitable in such formations. In granitic terrain, intermittent presence of *dykes* acts as barriers to groundwater movement. Dykes can be either magmatic or sedimentary in origin. Magmatic dykes form when magma intrudes into a crack and then crystallizes as sheet intrusion, either cutting across layers of rock or through the un-layered mass of rock. Dykes generally have high specific gravity (of about 3.0) and low porosity. The presence of dykes in an area is a negative indicator of groundwater. These act as subsurface dams for the lateral movement of groundwater. On the upstream side of the dyke, high yielding wells can be developed, whereas, there may be no groundwater on the downstream side. Rectangular shape of open wells is more preferred over the circular shape in the fissured or fractured rock terrains. The purpose is to expose the fractured rocks so that more water can be trapped into the well. The longer side of the well should be perpendicular to the trend of the fracture system.

2.10 Open wells in unconsolidated formations

The depth of the open wells in unconsolidated formations is usually kept about 7-10 m below the dry season static water level of the area. The diameter of the wells usually ranges between 1.5-4.5 m. To prevent cave-in of the well wall, bricks, stones, or RCC rings are used.

2.11 Tube well design

Pumping well design aims at its performance, service life, and cost. The location of the well and quantity of water required are two important criteria, which influence the well design. Usually, a service life of 10-15 years is considered for a well. Deep tube wells have two main elements *viz. housing/casing pipe* and *intake section*. The inside diameter of the well casing should be 5 to 10 cm larger than the maximum outside diameter of the pump and pump column. Well casing material should be selected based on groundwater quality, required strength, and cost. Incrustation is not a problem for well casing; however, it should be given proper treatment against acid corrosion, if groundwater is acidic in nature. The well yield depends upon the diameter of screen, but the two are not directly related. Keeping all other parameters same, doubling the screen diameter of a water table well increase the yield only by 11% (Linsely and Franzini, 1979); whereas, this increase is only 7% for a well installed in confined aquifer (Ahrens, 1958).

The minimum thickness of the pipe required for cutting threads on it is about 2.1 mm. This much thickness is considered additional to the design size.

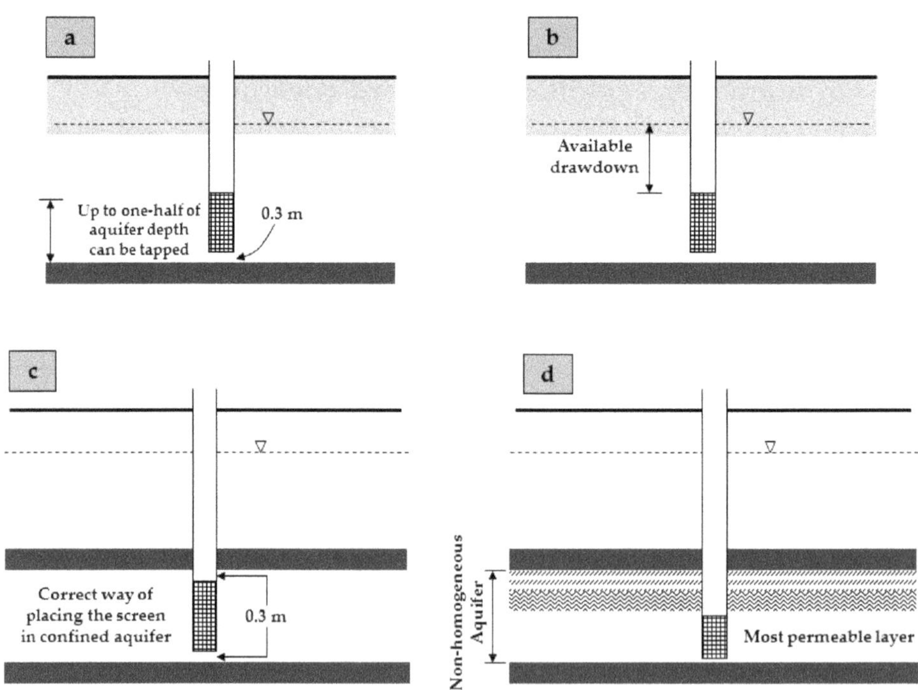

Figure 2.22: Placement of screen in the unconfined,confined, and semi-confined aquifer.

While installing the tube well in a homogeneous unconfined aquifer, bottom *one-third to half thickness* of the aquifer is tapped (with base of the screen at least 0.3 m above the impervious layer) (Figure 2.22a).

The purpose is to provide the maximum possible available drawdown. *Available drawdown* is the difference between the static water level and the top of the well screen (Figure 2.22b). Tapping of the homogeneous confined/semi-confined aquifer depends upon its thickness. In case aquifer is enormously thick, the screen may be placed in its upper central portion. Otherwise, confined aquifer's central 70-80% thickness should be tapped. For safety against the wrong placement of screen in the aquifer, it is always advisable to place blind pipe of length at least 0.3 m (or ~ 1 foot) at the top and bottom of the aquifer (Figure 2.22c). In case of non-homogeneous artesian aquifer, the most permeable layers are tapped (Figure 2.22d). However, in case of water table wells, the lower portions of the most permeable strata are tapped.

When the cone of depression of one well is overlapped by the cone of depression(s) of one or more adjacent wells, all are said to *interfere* with each other and as a result the yield of all the wells is reduced depending upon the overlapping. If the discharge rate of all the wells is known, then the drawdown at any point within the well field can be evaluated using the *principle of superposition*. According to which the drawdown at any point will be equal to the summation of drawdowns caused by each well individually. Mathematically, the formula can be written as

$$s_T = s_1 + s_2 + s_3 + \cdots + s_n \tag{2.49}$$

In Eq. (2.49), S_T and S_n represent the total drawdown and drawdown due to n^{th} well at the point under consideration, respectively. While pumping water from consolidated formations, screens are not required. However, for withdrawing water from unconsolidated formations, screens are required. Screens stabilize the borehole, prevent sand movement, and allow water movement from formation into well. The design parameters of a well screen include its diameter, length, total open area, slot size, and slot arrangement. Slot is an opening in the pipe for entry of water. The pipe, in which the slots are cut in certain pattern without compromising with its structural strength, is known as *slotted pipe*. Based on particle size distribution of aquifer, safe limit of entrance velocity into the well screen vary considerably. According to Bureau of Indian standards (BIS, 2000), the *entrance velocity through the screen* openings should be less than the permissible entrance velocity of 0.03 m/s. Bennison (1947) recommended that the entrance velocity through the screen openings must lie from 3.0 to 7.5 cm/s. However, Linsley and Franzini (1979) have proposed that safe entrance velocity may range up to 15 cm/s. The variation in the proposed velocity through well screen is due to different formations.

Length of the screen is governed by the aquifer thickness. For non-gravel pack wells, Walker (1974) has proposed a relation for evaluating length of the screen as

$$L_s = \frac{Q_0}{A_0 V_e} \tag{2.50}$$

In Eq. (2.50), L_s is the screen length (m); Q_0 is the maximum expected discharge capacity of well (m^3/min.); A_0 is the effective open area per meter length of the screen (m^2/m); and V_e is the entrance velocity of water into the well screen.

Diameter of the well screen depends upon the designed yield of the tube well. While evaluating the open area of the screen, allowance is given for clogging, possible coverage by gravel pack material, and possible incrustation. On an average, around 50% of open area of the screen is blocked. In general, the percent open area of the screen is kept in the range of 15 to 22%. For attaining better transmissivity, the percent open area should be as high as possible. However, there are studies and even recommendations that without compromising with the structural strength of the screen, its open area can be kept up to 36% (Singh and Shakya, 1989).

Decision of *screen's slot size* selection is based on the grain size distribution of aquifer material. In a gravel pack well, the slot size is so selected that it should retain 90% of the gravel pack material (BIS, 2000). In case of non-gravel pack wells, the screen should retain 30 to 60% of the formation material. The screens with slot sizes of 1.0, 1.6, and 3.2 mm are available in market.

In general, the shape of the slot is such that it widens inwards (Figure 2.23). The slots should be cut in a pattern to achieve even distribution of flow all over the periphery of pipe.

The material of the well screen should be corrosion resistant and have enough structural strength to prevent collapse. Generally, used corrosion resistant materials are copper, bronze, nickel, and stainless steel. Bureau of Indian Standards (BIS, 2000) has classified the screens in various categories *viz.* (i) Plain slotted pipes – Type A, (ii) Bridge slotted pipes – Type B, (iii) Mesh wrapped screens – Type C, (iv) Cage type wire-wound screens – Type D, (v) Pre-packed resin bounded, gravel screens – Type E, (vi) Brass screens – Type F. As per the recommendations, slots may be made in groups of 2 to 5 having length of 75 mm (Figure 2.24). The recommended

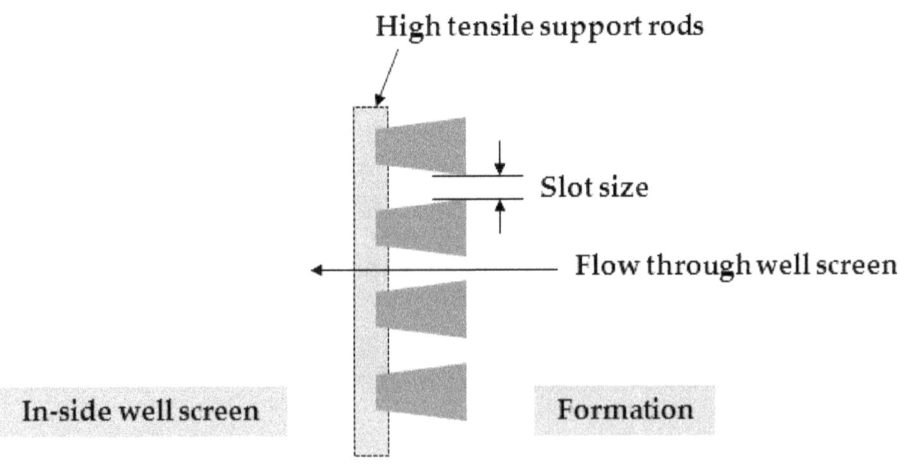

Figure 2.23: Schematic sketch showing the geometry of well screen.

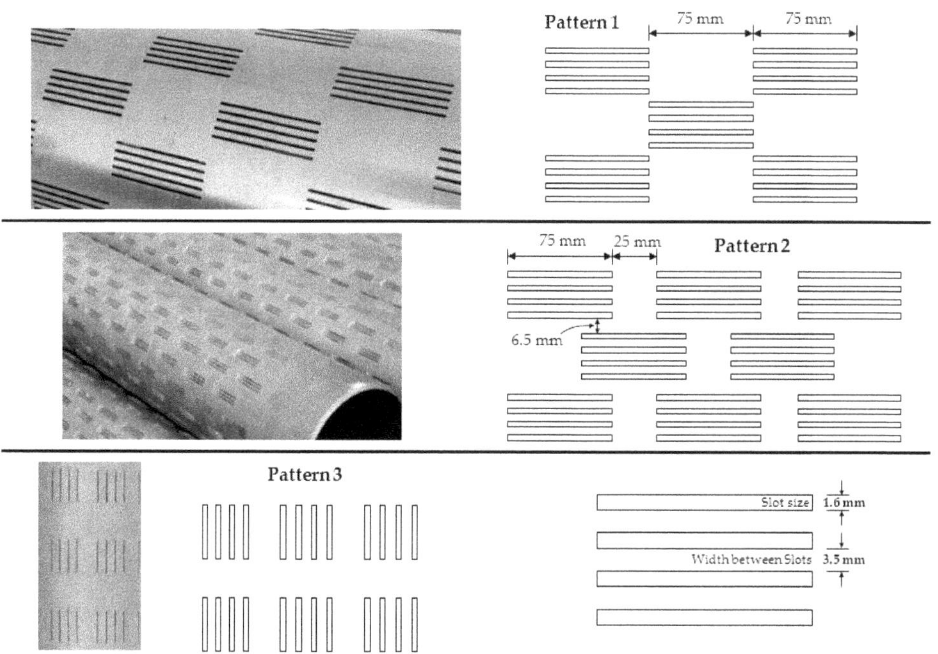

Figure 2.24: Schematic sketches and photographs showing different slot patterns on well screens.

spacing between the slots is 3.5 mm. Mild or low-carbon steel pipes are used for making mesh wrapped screens sections of length 3 to 6 m. For limiting the flow resistance through screen, the diameter of perforations should not be less than 12 mm. Based upon the diameter and thickness of pipe, the distance between two adjacent perforations should lie between 25 and 40 mm (BIS, 2000).

A copper mesh with minimum thickness of 0.710 mm is wrapped on the perforated pipe by placing spacers (6 mm wide and 3 mm thick) at 75 mm apart from each other. Most commonly, this strainer is known as *agricultural strainer* (Figure 2.25a). Pre-packed gravel screens are also available in market. With the help of resin-type adhesive material, gravel pack is pasted on the perforated pipe (BIS, 2000). Depending upon the diameter of the perforated pipe, the thickness of the gravel pack may vary from 10 to 15 mm (Figure 2.25b). Leaded brass strips conforming to Indian standards (BIS, 1981) are used to manufacture brass screens. Depending upon the diameter and depth of well, thickness of brass sheet required for manufacturing pipe varies from 3.2 to 5.4 mm. Other than the above-mentioned commercial screens, farmers are manufacturing coir and bamboo screens at local level. Coir screens are manufactured by wrapping coir on cylindrical mild steel frame (Figure 2.25c). The complete surface of the coir acts as the screen. However, the manufacturing cost of screen is very less but its service life is limited to 5 – 10 years. Bamboo strainer is similar to coir strainer except the cylindrical frame, which is manufactured with bamboo strips (Figure 2.25d). Michael (1997) has reported that bamboo screen was originally developed in *Saharsa* district of Bihar (India).

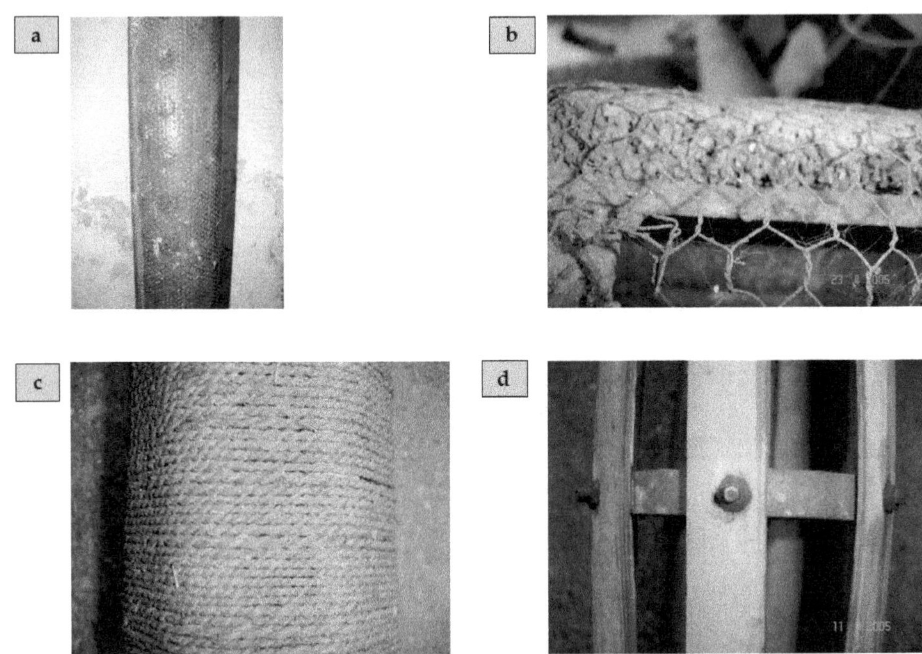

Figure 2.25: Photographs showing (a) agricultural strainer, (b) pre-packed gravel filter screen, (c) coir screen, and (d) bamboo screen.

2.11.1 *Gravel packing* – The placement of graded sand, gravel, or fibrous material around a well screen to prevent infiltration of fine materials is known as pack material. More specifically, if only gravels are used for the purpose, the term is known as *gravel packing*. The placement of material from outside artificially is known as *artificial gravel packing*. Artificial gravel packing is also known as *gravel shrouding* or *gravel filter*. However, when the finer fraction of the aquifer formation is pumped through screen leaving the coarser content around the well screen is known as *natural gravel packing*.

Sieve analysis of the aquifer material is performed to know the characteristics of water bearing formations and the results so obtained provide the basis for well screen specifications and gravel pack design. In India, sieve analysis is conformal to Indian Standards (BIS, 1962).

2.11.2 *Effective size* – The term *effective size* (D_{10} or d_{10}) is defined as the size of sieve through which only 10% of the material is passed and 90% of the material is retained.

2.11.3 *Uniformity coefficient* – The term *uniformity coefficient* was proposed by Hazen, which expresses the degree of assortment of water bearing sand as an indicator of porosity. It is the ratio of the sieve size, which passes 60% of the material to the sieve size, which passes 10% of the material. Mathematically, it can be expressed as $C_u = \frac{D_{60}}{D_{10}}$. For one size grains (i.e. *complete assortment*), the value of $C_u = 1$ and for

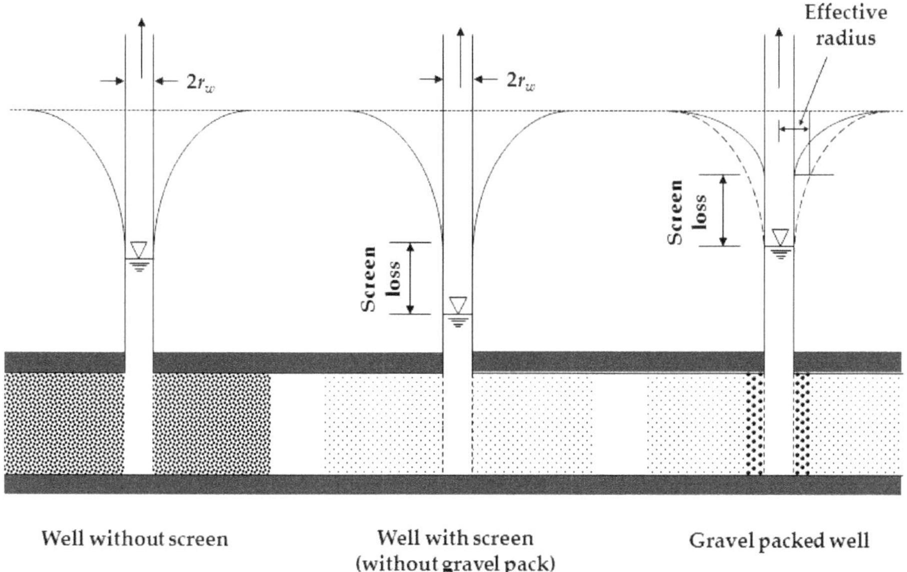

Figure 2.26: Effect on pumping water level and drawdown by the uncased, screened, and gravel packed well.

even-grained sand it ranges from 2 – 3. For heterogeneous sand, the value will be even high. In general, the formations with an effective size of 0.25 mm and uniformity coefficient of 2 or more do not require gravel packing. Laminated and/or fine sand formations require gravel packing.

2.11.4 *Laminated formation* – The alternate combination(s) of thin layers of fine, medium, and coarse sand in a formation is known as *laminated formation.*

2.11.5 *Effective radius* – If well is fully developed, gravel packing (either natural or artificial) increases the *effective radius* of the well (Figure 2.26). Consequently, the entrance velocity through screen declines, which further reduces *incrustations*. The gravels should be clean, well-rounded, uniform, smooth, water-insoluble, and nonacid soluble. The soft or water soluble materials are not used as gravel packs (e.g. anhydrous $CaSO_4$, Gypsum, or shale etc.). In no case, the particle size of the gravel pack should be more than 13 mm (BIS, 2000).

2.11.6 *Pack-aquifer ratio* – Design criteria for artificial gravel pack is expressed in terms of *gravel pack ratio* or *pack-aquifer ratio* (P-A ratio) which may be expressed as

$$Pack-aquifer\ ratio = \frac{Average\ size\ of\ gravel\ pack\ material}{Average\ size\ of\ aquifer\ material} \quad or$$
$$P-A\ ratio = \frac{50\%\ size\ of\ gravel\ pack}{50\%\ size\ of\ aquifer} \tag{2.51}$$

Based upon the minimum head loss and sand movement through the gravel pack, the design criteria recommended by Bureau of Indian Standards (BIS, 2000) is presented in Table 2.1.

Table 2.1: Design criteria for gravel pack.

Pack	Pack-aquifer ratio
Uniform aquifer with uniform gravel pack	9 : 12.5
Non-uniform aquifer with uniform gravel pack	11 : 15.5

If a well is tapping more than one formations, then the gravel pack designed for the formation with finest aquifer material is provided for all the formations subject to the condition that

$$\frac{50\%\ size\ of\ the\ coarsest\ formation}{50\%\ size\ of\ the\ finest\ formation} < 4.0 \quad (2.52)$$

The slot size of the screen for a gravel pack well depends upon the size of designed gravel pack and it should be such that it retains 90% of gravel particles. Laboratory studies show that only a few mm thick gravel pack layer can efficiently retain the formation particles regardless of the flow velocity through the pack, which has a tendency to carry formation particles through it (Johnson, 1966). However, the placement of such a thin layer of pack is rather difficult. Normally, a gravel pack of

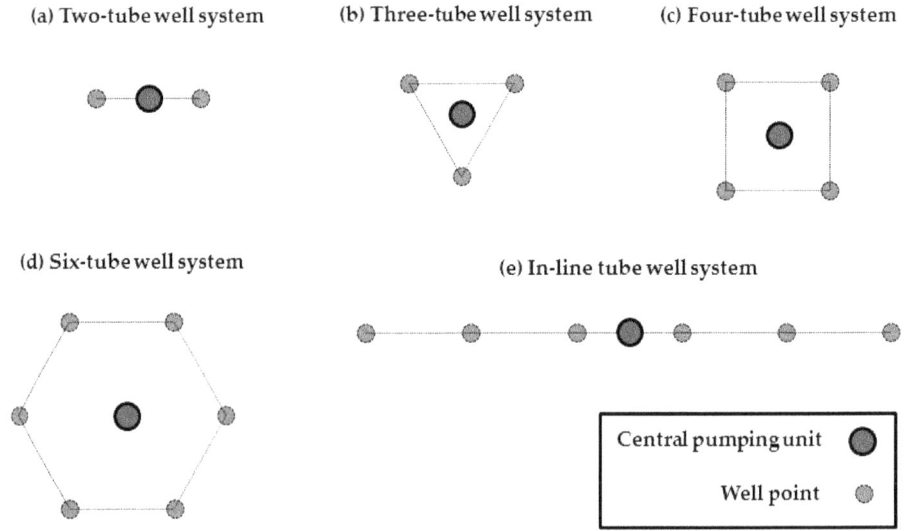

Figure 2.27: Different geometrical patterns in which multiple well points system can be installed.

thickness ranging from 50 to 150 mm is adapted (ASAE, 2005). The upper limit of the gravel pack thickness is decided by the size of bore and casing pipe and generally not exceeds 20 cm. Most importantly, the well yield does not depend on the gravel pack thickness; but it becomes difficult to develop well with thick gravel pack.

2.12 Multiple well points system

When more than one well is attached to a single pumping unit, the system is known as *multiple well points system* (MWPS). Depending upon the number of wells, different geometrical patterns can be thought of (Figure 2.27). In general, MWPSs are used for lowering groundwater level (either for civil construction or for drainage) and for skimming floating freshwater from fresh-brackish/-saline aquifers. Therefore, the spacing between the individual wells depends upon the purpose of its use. Water table can be lowered to greater extent by installing the well points close enough so that these could interfere. In contrast, the MWPSs installed for skimming freshwater are designed in such a way that their drawdown curves should not interfere otherwise pumped water quality may deteriorate.

3

Well Construction

3.1 Introduction

A water well is a hole (either vertical, or horizontal, otherwise combination of two) excavated up to water bearing formations for bringing water to the surface. Wells are not only used for abstracting water from underground formations, but are also used to inject depending on the specific purpose. Groundwater recharging, storage of freshwater in saline aquifers, disposal of treated or untreated wastewater in brackish aquifers, deep well injection of liquid hazardous wastes, the formation of barriers to restrict seawater intrusion, and drainage of inundated lands are few examples of injection through wells (Vashisht and Shakya, 2015). Wells can be constructed by many methods and the selection of a particular method depends upon the purpose of well, the quantity of water required, groundwater depth, geology of underground formations, and economic factors. Shallow wells can be installed by digging, driving, jetting, or boring by bailer. For installing wells in deeper formations and in hard rock areas, percussion, rotary (with drilling mud), reverse circulation rotary (with muddy water), air rotary, or air percussion rotary drilling methods are used. On the completion of drilling, well is installed, gravel packed (if required), developed for optimum yield, and tested.

Description regarding the drilling equipment, well installation, well development, and general tube well troubles are covered in this chapter.

3.2 General recommendations

The following steps should be kept in view while constructing a new well.

- An accurate map of the proposed location of irrigation well should be drawn.
- Initial investigations should be carried out for the presence of any obstruction on the site such as high-tension power line and underground cables or pipelines.
- Documentation should be complete (specifically the electric connection).
- It is the liability of the owner to provide healthy and safe working environment for the workers. In few countries, there is a provision of insurance also.
- The drilling site should be easily accessible and enough working space should be there.
- A source of good quality water should be present near to the drilling site, and if not, proper arrangements should be made in priori.
- It is the responsibility of the driller or supervisor to collect the following information.
  a) Type and diameter of casing
  b) Length, diameter, and position of the screen
  c) Gravel pack particle size analysis
  d) Position of seals or stabilizer blocks
  e) Well bore diameter

3.3 Well construction procedure

3.3.1 *Open wells* – Open wells in un-consolidated formations are excavated manually using pickaxe and shovel. In semi-consolidated formations, these are excavated with dragline excavators using clamshell or orange-peel type buckets.

The yield of open wells in hard rock and semi-consolidated areas can be increased by boring suitable length lateral tunnels (also known as *revitalization holes*) below the water table. In semi-consolidated and hard rock areas, mechanically powered rock drilling equipment is used which is capable of drilling holes of size 4 – 5 cm in diameter and 25 – 30 m long. In alluvial and other unconsolidated formations, radial filter pipes are installed as unlined bores may collapse.

Tube wells can be constructed by *driving*, *jetting*, or *drilling* methods. Driven and jetted tube wells are limited to shallow depths up to 15 m.

3.3.2 *Driven tube well* – As the name indicates, *driven tube wells* are forced into the aquifer with the help of suitable weights (i.e. wooden maul, drop hammer etc.). To ease in driving action, a steel cone is fixed at the bottom of the screen. For protecting the top of pipe (threaded portion) while hammering, a drive head (or drive cap) is fixed on it. While driving the pipe it is kept full of water, which escapes as soon as it encounters the porous formation. Usually, the size of driven tube wells ranges between 3 – 7.5 cm; however, if the well has to pump water from a depth more than 7 m the size of the pipe should not be less than 5 cm. Driven tube wells may yield in the range of 100 – 250 m^3/d.

3.3.3 *Jetted tube well* – These tube wells are constructed using the cutting action of downward-directed water jet. To prevent cave-in of the borehole a casing pipe is lowered as the cutting proceeds. The casing pipe also assist in conveying cuttings along with water out of well. To ensure a straight hole, it is advisable to keep on rotating the drill pipe slowly as the drilling proceeds. After placing the screen and blind pipes in the borehole, casing pipe is pulled out.

3.4 Drilling equipment

The different drilling equipment are classified as

- Percussion drills (also known as cable-tool drills)
- Rotary drills
    a) Direct rotary
    b) Reverse circulation rotary
    c) Air rotary
- Hammer drills
- Core drills

All the above-mentioned drills can be further classified into light, medium, and heavy.

3.4.1 *Percussion drills (Cable-tool method)* – In this method, the drilling is performed by the continuous operation of dropping and lifting of chisel edged bit. It makes 20 – 40 strokes per minute, which may range from 0.40 – 1.00 m in length. The length of the bit ranges from 1 – 3 m and it may weigh up to 1500 kg. The casing pipe guides the dropping movement of the bit. To avoid the damage that may cause to the first section of casing pipe while driving through formation, a *drive shoe* with a beveled edge is fastened at its bottom. There is a need to add water from outside if the bore has not encountered any water bearing formation. It is required to form the paste of cuttings so that the friction it caused to falling bit can be reduced. After every 1 – 2 m drilling, cuttings are bailed out. In consolidated formations, this method is capable of drilling bores of size ranging from 8 – 60 cm and up to a depth of 600 m. Bailer (or sand bucket) is used to remove drill cuttings from the bore. Its specifications are shown in Figure 3.1. The sudden rise or decline of water level during the percussion drilling indicates the presence of permeable bed. While drilling deeper bores (specifically in hard rock areas), it is necessary to maintain the alignment. If there is any tendency of the bore to misalign due to the presence of any obstruction, it can be corrected by detonating controlled explosion at the bottom. In Table 3.1, advantages and disadvantages of the cable-tool method are described.

Table 3.1: Advantages and disadvantages of the cable-tool method.

Advantages	Disadvantages
The equipment is simple in design, is rugged, and easy to maintain. Detection of thin and weak water bearing zones. Suitable method in arid and semi-arid regions as very less water is required for drilling. For preparing well log, formation samples can be accurately procured. It is possible to test the quantity and quality of water in each stratum as drilling proceeds.	Drilling rate is slow. Drilling depth is limited. It is necessary to lower the casing pipe coincidentally with drilling in unconsolidated formations. It is not suitable for drilling in dune sand, quick sand, and unconsolidated river gravels as the pounding of drill, non-sealing of drilled bore, and low-level of water promotes slumping and caving-in of such formations.

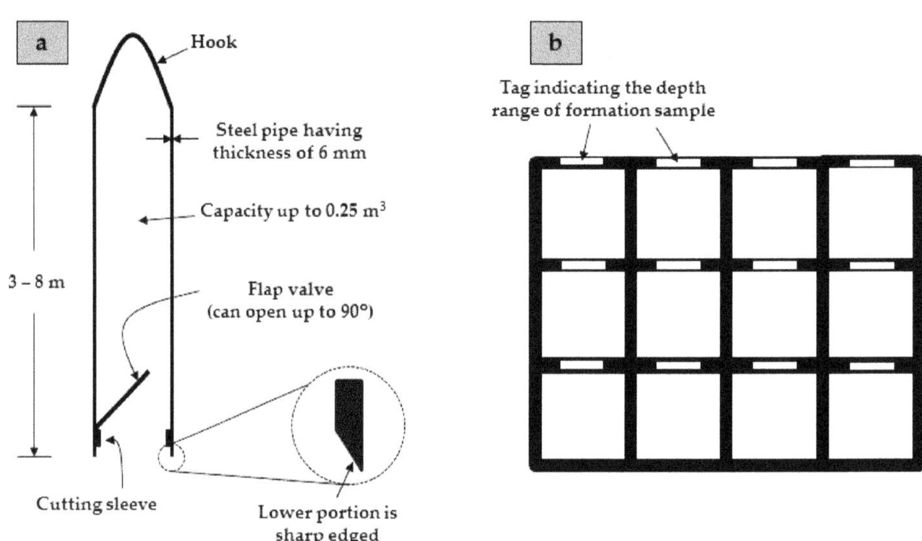

Figure 3.1: Schematic sketches of (a) bailer and (b) formation sampling box.

3.4.2 *Rotary drilling* – It involves the drilling of wells with the aid of rotating bit to which downward force is applied. The bit is attached to the hollow stem through which a drilling fluid (usually a mixture of clay and water) is conveyed-to or pumped from the drilling point. The *drill collar* is used to hold the bit to the lower end of the drill pipe. The method is rapid and specifically adapted to drill deep and large size bores. The method is unsuitable for drilling in boulders, other hard rocks, slanted, and fissured formations (i.e. consolidated formations). Casing is not required during the drilling process. Because of this advantage, the test holes can be abandoned at a minimum expense. In extremely permeable lost circulation zones, rotary method is unsuitable as drilling fluid lost largly in such formations. The method is also not suitable for drilling unconsolidated formations having boulders and cobbles, as it is difficult for the circulating fluid to take these out of bore.

3.4.2.1 *Direct rotary drilling* – In this method, drilling mud (or drilling fluid) is forced down through the hollow drill stem and out through the nozzles in the bit.

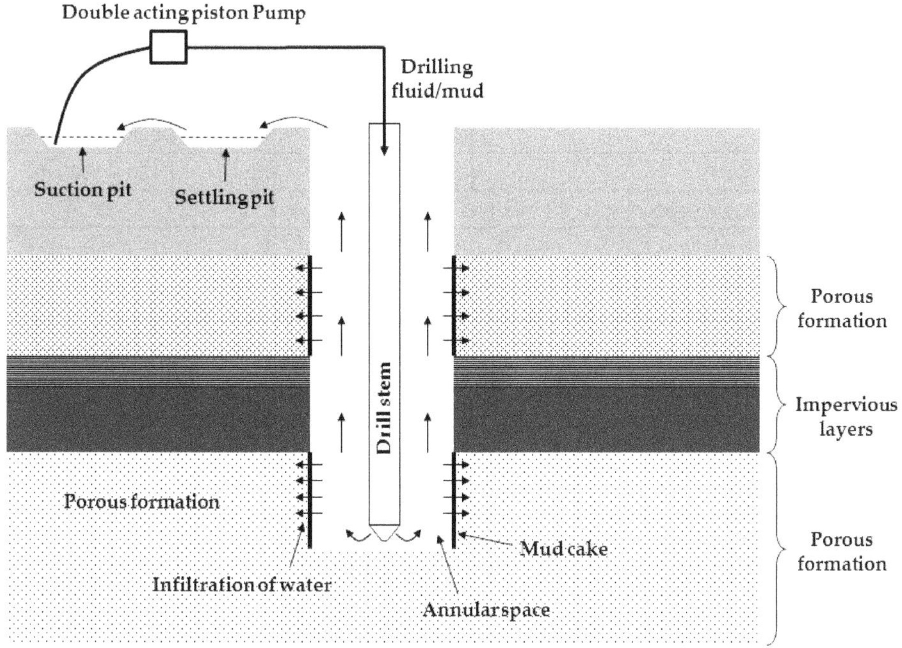

Figure 3.2: Schematic sketch showing the operation of direct rotary drilling.

Materials loosened by the bit are picked up by the drilling fluid, carried it to the surface through the annular space between drilling stem, and bore hole. Ideally, the velocity of flow through the annular space ranges between 40 – 60 m/min. While passing through the annular space, porous bore wall acts as filtration media, through which water from the drilling mud is escaped to the adjoining formations by infiltration (Figure 3.2). As a result, a clay lining is developed on the bore wall, which is known as *mud cake*. The sealing of wall further reduces the loss of water into permeable formations. The drilling mud is channelized to settling pit from where it over-flows to suction pit. As long as the hydrostatic pressure of the fluid exceeds the earth pressures and any artesian pressure in the aquifer which tend to collapse the hole, the hole will remain open. A pump capable of forcing drilling fluid with a pressure of 9 kg/cm^2 or more is used. Usually, double acting piston type pump is preferred. Drilling mud is prepared by mixing bentonite clay and various other organic additives in water. The quality of the drilling mud is judged based upon the parameters which include its density, viscosity, gel strength, filtration property, and percentage of suspended solids. The drilling speed is usually kept in the range of 30 – 150 rpm. A bit pressure of 100 – 200 kg/cm^2 per cm diameter of the bit is maintained while drilling.

Two types of bits are used for drilling *viz.* (i) *drag bit*, and (ii) *rolling cutter type bit*. Rolling cutter type bit is also known as *rock bit*. The drag bit produces the shearing and scraping action and is therefore, suitable in soft and unconsolidated formations.

Table 3.2: Capacity range of rotary rigs.

Particular	Light	Medium	Heavy
Maximum hook capacity	900 kg	13000 – 18000 kg	22000 – 45000 kg
Size of drill pipe	6 cm	6 – 7 cm	9 cm
Maximum drilling depth	60 – 140 m	450 – 700 m	750 – 1000 m
Average drilling speed (depending on the size and depth of bore, type of formation) = 5 – 30 m/d			

A two-way bit or fish tail bit (i.e. because of its resemblance with fish tail), a three-way bit, or a six-way bit (or pilot bit) are various forms of drag bit. According to the name, the mentioned bits have the same number of cutting blades. Pilot bit (or six-way bit) is more suited to shattered formations and cemented sands. For drilling in the hard formations, roller type of bit is adapted, which exerts crushing and chipping action. Cone and cross-roller bits are two major types of roller bits. Cone roller bit contains three cutters, whereas cross-roller has four cutters.

Shearing of drill pipe, unthreading of the couplings and accidental drop of drill pipe in the hole are the frequently occurring incidences while drilling well bores. The tools used to recover these are known as *fishing tools*, few of them can be named as *tapered tap*, *die overshot*, and *circulating slip overshot*. The performance of tapered tap and die overshot is more effective, if run immediately after the failure. The specifications of the direct rotary drilling based on the capacity are presented in the Table 3.2.

3.4.2.2 *Reverse circulation rotary drill* – In this method, drilling fluid is circulated in reverse to the direction of that of the direct rotary drilling method (Figure 3.3). Clear water is used as drilling fluid in this case which enters the drilling process through the annular space between drilled bore and drill stem and is allowed to mix with the drill cuttings. Clay can be added in the water if cave-in starts. The complete blend is recovered to the surface by a centrifugal pump through the drill stem with a velocity of flow exceeding 2 m/s. The bore is not cased during the drilling process as the chances of cave-in are prevented by the hydrostatic pressure of the water column subject to the condition that water level must be kept at ground level at all times.

Moreover, a water head of at least 3 – 4 m is required for promoting the suspended particles in water to seal the bore wall. The erosion of the bore wall is restricted by limiting the flow velocity of drilling fluid that is possible by increasing the diameter of well (with a minimum of 0.40 m). That is why this method is suitable for drilling large size bores, generally in the range of 0.40 – 1.8 m (bit size). Depending upon the size and depth of hole and until the gravel packing of the well, water supply in the range of 3 – 9 l/s is required. The reverse circulating rotary drill differs from the direct rotary drill in the following ways

- For drilling bores up to a depth of 300 m or more, airlift method is preferred.
- The size of the drill stem is more than the direct rotary drill of same capacity. *Flange joints* are used to attach different sections of the drill stem.
- Centrifugal pump with non-clog impeller is used in drilling.

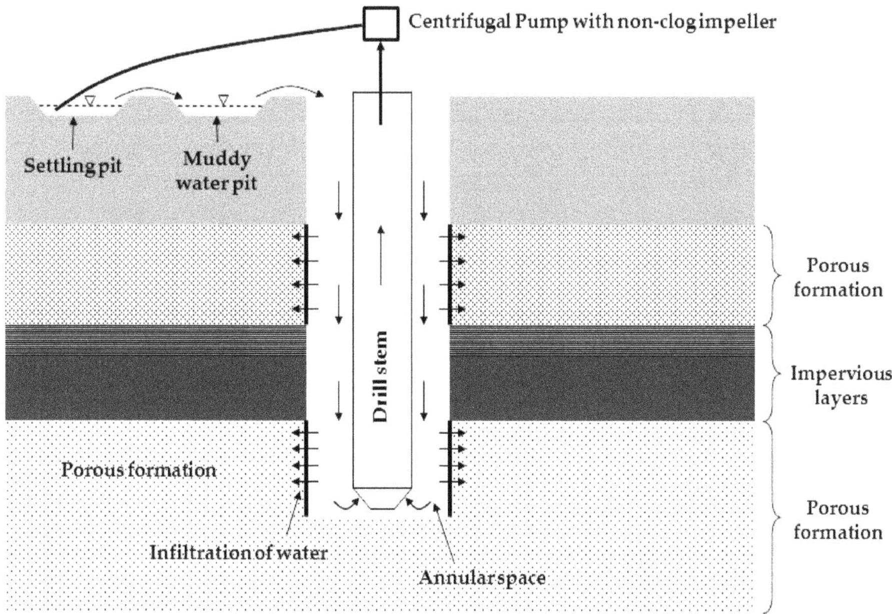

Figure 3.3: Schematic sketch showing the operation of reverse circulation rotary drilling.

- The drilling fluid is termed as *'muddy water'* against the term *'drilling mud'* in case of direct rotary drilling method.
- In comparison to direct rotary drilling, velocity of drilling fluid (i.e. water) in the annular space is low and does not erode the bore wall.
- The method requires about 5 times more water than the direct rotary drilling method; hence, restricted to locations where enough supply water is available.
- Since, clear water is used as drilling fluid there is no mud invasion in the aquifer formation(s). Therefore, well development requires very less time in comparison to direct rotary drilling.

The method is most suitable for drilling very large size holes in unconsolidated formations. The reverse circulation drilling method is not suitable in the following cases

- In formations, where cave-in is prominent.
- Exploratory test drilling.

3.4.3 *Air rotary drilling* – When the compressed air instead of drilling mud is used in rotary drilling equipment, it is named as *air rotary drilling*. In consolidated formations, where clay lining of the drilled bore is not required, drilling can be performed rapidly and conveniently. Air circulation is similar to the drilling mud circulation in direct rotary drilling technique. To remove the cuttings from the hole, an air velocity of 900 m/min. in the annular space is adequate. However, depending upon the formation and size of bore, the recommended air velocity may range from 600 – 1500 m/min. Sometimes, it become necessary to use foams and other air additives to drill large size bores. The major advantage of the air rotary drilling is its capability to drill through fissured rock formations with little or no requirement of water.

3.4.4 *Air rotary percussion drilling* – For drilling bores in hard-rock formations, *rotary-percussion drilling* technique is used. Since, air is used as drilling fluid it is sometimes also known as air rotary percussion drilling. This method combines the percussion action of cable-tool method and rotary action of rotary drilling. The tool used in the rotary-percussion drilling is known as *down-the-hole air hammer*. The bit of the tool is formed by inserting tungsten carbide in alloy steel (Figure 3.4).

A pressure of 7 – 17 kg/cm^2 is required for operating the pneumatic hammer. The hammer delivers 10 – 15 impacts per second to the rock. A penetration rate of 0.3 m/min. can be achieved depending upon the formation. For smooth drilling, the hammer should be in continuous contact with the face of the rock at a pull down pressure in the range of 30 – 150 kg per cm diameter of the bit. With increase in the depth of hole, pull down pressure may be compensated with the weight of drill pipe added to drill stem. The rotational speed of the drill bit

Figure 3.4: Photograph of the bit with tungsten carbide buttons.

ranges from 15 – 50 rpm. Rotation rate will be lower for hard or more abrasive rock. The method is most suitable for drilling small diameter wells in hard rock formations.

3.5 Common drilling problems

3.5.1 *Lost circulation* – An excessive loss of drilling fluid through heavily porous medium or through formation having joints and fissures is termed as *lost circulation* (Figure 3.5). Altering the circulating fluid properties may help to get rid of this problem.

3.5.2 *Crooked holes* – The undue eccentricity during drill stem rotation is the basic cause of *crooked holes* (Figure 3.6). These can be avoided by (i) keeping the size of the drill collar large enough in diameter to hold the pipe and bit in the center of hole, (ii) avoiding excessive bit pressures, (iii) keeping the diameter of the hole less than twice the diameter of the drill pipe, and (iv) keeping the bit rotation speed more than 150 rpm.

3.5.3 *Bailing up* – The accumulation of formation material such as clay, sticky shale around the bit and drill collar is known as *bailing up*. The high rate of penetration combined with the insufficient speed of bit rotation (to mix the drill cuttings) is its normal cause. To remove this, the drill pipe is spudded frequently by raising and dropping it within a vertical distance of 1 – 1.5 m while the pipe is being rotated rapidly.

3.5.4 *Stuck up drill pipe* – The sudden reversal of circulation fluid due to lost circulation in the hole deposits the suspended drill cuttings around the drill pipe

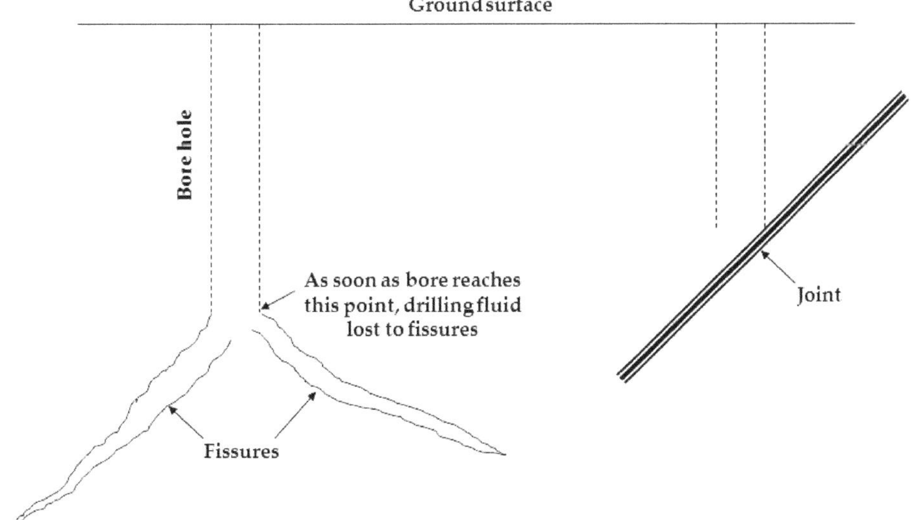

Figure 3.5: Schematic sketches showing the presence of fissures and joints, which are responsible for lost circulations.

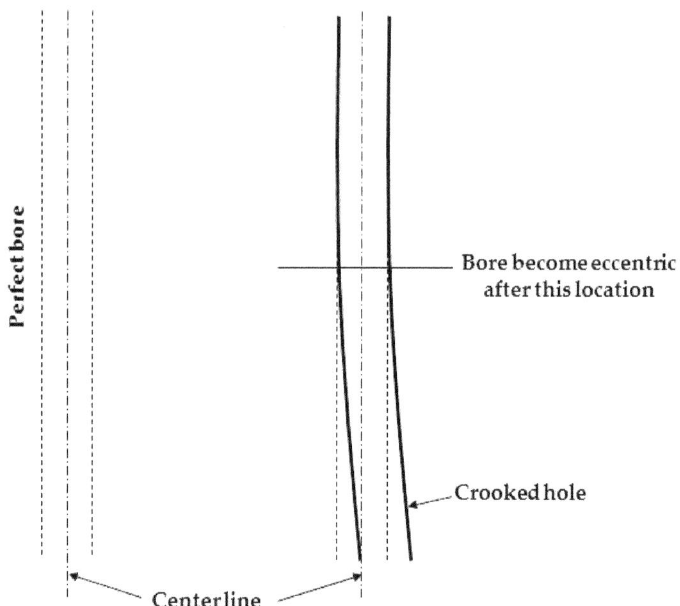

Figure 3.6: Schematic sketch showing the crooked hole.

which stick to it very tightly and *stuck up drill pipe*. It is only with the jacks the drill pipe is pulled.

3.6 Well installation

3.6.1 *Bail plug* – The *bail plug* is a cap used to close the bottom of 1 – 2 m long blind pipe fixed at the bottom of the well assembly (Figure 3.7). An eye is fixed on the bail plug to hook whenever it is necessary to pull out the well assembly. During well development, it collects the sand particles that enter the borehole.

3.6.2 *Sealing of brackish/saline aquifer horizon – Drill stem* method is used to demarcate the saline horizons. The density of the cement slurry used for sealing the unwanted formations varies with its depth of application. The desired density of the slurry is achieved by adding gel, expanded perlite, hydrocarbons, barite, $CaCl_2$, and $NaCl$.

3.7 Well development

After well construction, it is necessary to develop the well to obtain its maximum capacity for a given drawdown. Usually, it is urgent to develop all gravel packed and screened wells except when screen is formed of fine wire mesh or coir or other closely-knit filters located in a highly permeable formation (Michael, 1997). In the process, fine materials from the formation (near the well screen) are removed, thereby, opening the passages so that the water can enter the well more freely. Principally, the formation particles are severely agitated and allowed to rearrange by reversing the flow through screen. This process breaks the bridging between particles and allows the finer particles to enter the well. The *well development* is performed to (i) remove any clogging of aquifer, (ii) increase the permeability of

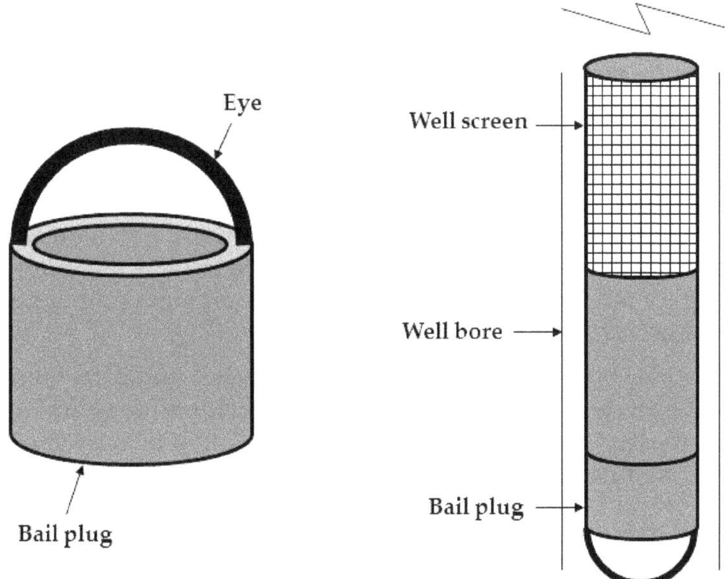

Figure 3.7: Schematic sketches of the bail plug and the way it is installed in the well.

the aquifer in the vicinity of well screen, and (iii) stabilize the aquifer formation. The method used to develop a well should be selected based upon the geologic makeup of the aquifer.

3.7.1 *Surging or swabbing* – It is accomplished by up and down movement of a close-fitting *surge plunger* (or *surge block*) which alternately forces the water out into the surrounding formation and then back to well. Well development is started with light surging; however, the force of the surge increases as the development proceeds. Surge plunger is of two types: (i) a solid plunger, and (ii) in valve openings are provided and are generally used for lighter surging. Surge plunger should be 3 – 5 m below the water level in the well. In general, 2 – 3 days are required to complete the well development by surging.

3.7.2 *Bailing* – The method is used for the wells having diameter more than 20 cm. The line speed of the closely fit bailer should be between 2.5 to 5.0 m/s while hoisting and free-fall during its lowering (ASAE, 2005).

3.7.3 *Pumping* – This development method should be an additional or finishing development step when used in combination with other methods and is not recommended as singular method. This method is not very effective in every formation as the flow of water remains only in one-direction, which promotes bridging. Pumping is started at about $1/4^{th}$ to $1/5^{th}$ of the designed discharge rate of the well and is continued until water becomes clear. Pump is turned-off which created surge in the well. After some time, pump is again started. The process is repeated at 50, 75, 100, and 125% of the designed well yield (ASAE, 2005). While pumping at the rate of 125% of the designed well yield, it is categorized as *over-pumping*, which means to pump the well with excessive drawdown.

3.7.4 *Rawhiding* – It is a well development method in which the rapid changes in the head of the well are produced by starting and stopping the pump intermittently. Depending upon the time provided between two consecutive pumping operations, this method could be employed in two different ways. In the first way, the pump is operated at its fullest capacity to produce the greatest drawdown. After stopping the pump, well is allowed to recover its original static level. In the second way, first step is same (i.e. to produce the greatest drawdown and then stopped). However, pump is again started after short interval (i.e. not allowed to recover its original static level). Wearing of the pump due to sand pumping is a major disadvantage of this method. Under severe sand pumping conditions, *sandlocking of pump* may occur.

3.7.5 *Backwashing by pumping* – In this method, water under pressure is forced in well which eventually passes through the screen openings and agitates the formation particles. On stopping the backwashing, the flow reversal through aquifer accumulates the finer fraction of the formation in the well, which is subsequently bailed out.

3.7.6 *Compressed air* – In this method, compressed air is used to develop well by backwashing and surging method. In the backwashing method (also known as closed well method), compressed air is introduced in the well (through an airline in the

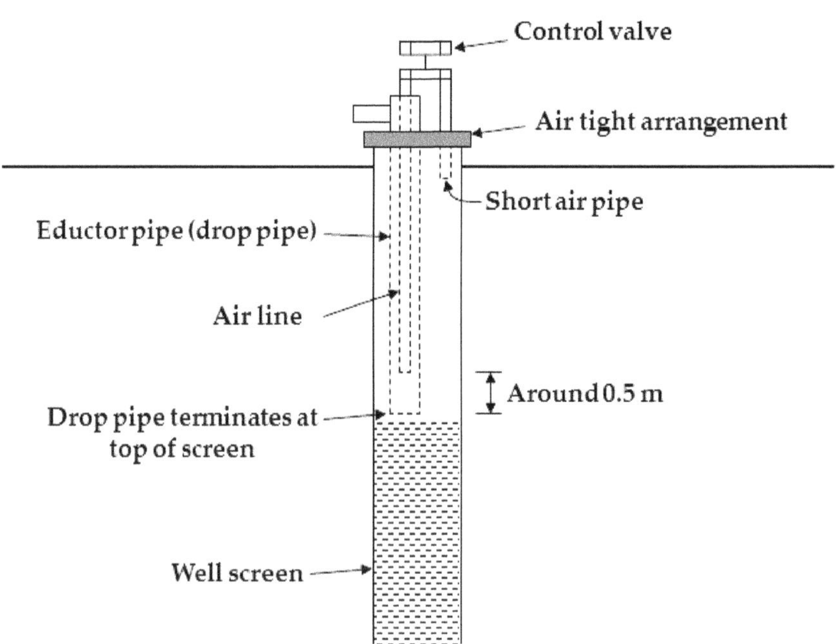

Figure 3.8: Schematic sketch showing well development by backwashing with compressed air.

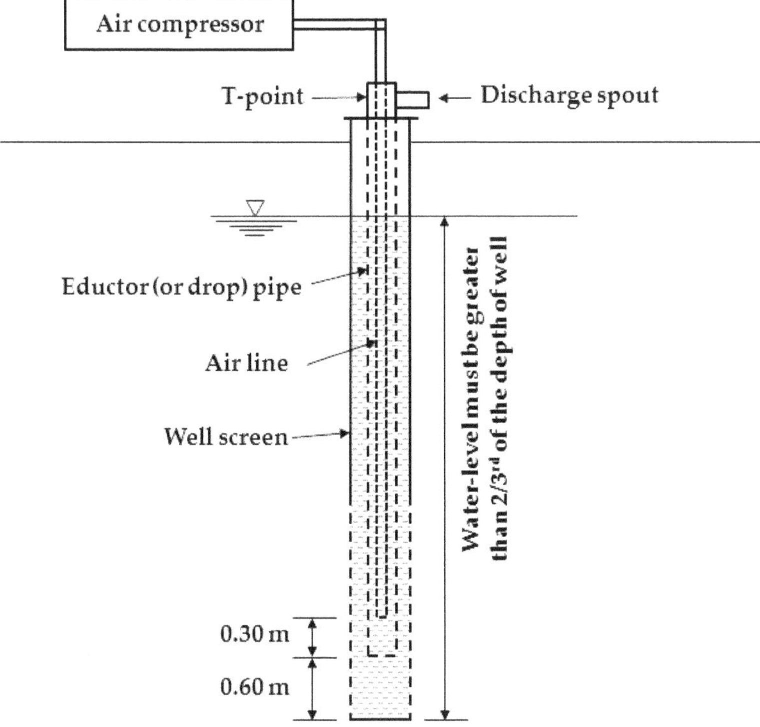

Figure 3.9: Schematic sketch showing well development by surging with compressed air (open well method).

cap used for closing the top of well) to force the water back out of the well through the screen and into the aquifer (Figure 3.8). With the release of air pressure, water flows back from aquifer to well and bring fine sand with it. The surging method by compressed air (also open well method) is limited to wells in which depth of water exceeds the $2/3^{rd}$ of the total depth of well. The surges are created in the screened section of the well by compressed air (Figure 3.9). Nowadays, a double-packer tool is utilized for selectively developing the well inlet.

3.7.7 *Jetting method* – The use of high-pressure water jets is an effective method of well development and is commonly known as *jetting method.* The jet points at the screen openings whose forceful action clears the screen and agitates the formation. The clay wall-cake deposited on the formation (while rotary drilling) can be removed very effectively.

3.7.8 *Use of dispersing agents* – Poly-phosphates has a property to disperse the clay particles. Dispersed clay particles can be removed more effectively by surging and backwashing methods. Commonly used chemicals are tetrasodium pyrophosphate ($Na_4P_2O_7$), sodium tripolyphosphate ($Na_5P_3O_{10}$), and sodium hexametaphosphate

$(NaPO_3)_6$. Use of 1 kg of chemical per 200 l of water can effectively disperse the clay particles. In limestone formations, acid may be used to enlarge openings and improve production.

3.8 Cavity well development

It means the development of a stabilized cavity below the clay roof after removing sand from the aquifer. Studies show that for a fully developed cavity, its circular area is much larger in comparison to its depth. It is preferred to use centrifugal pump for developing a cavity.

3.9 Disinfection

To avoid the entry of any bacteria during the well construction and development process, the drilling and development equipment should be disinfected with a chlorine solution. The chlorine concentration in the solution should be at least 100 mg/l.

3.10 Tube well troubles

3.10.1 *Pumped water is muddy and/or contains fine sand and silt* – Muddy water is due to wrong placement of screen in the aquifer. In case, the screen is placed against the confining/semi-confining layer, it starts dislodging confining layer particles. Presence of fine sand or silt may also be due to improper selection of screen slots, improper well development, or improper gravel pack design and placement.

3.10.2 *Low tube well discharge* – Improper design of well and/or gravel pack are the basic reason of low tube well discharge. However, lack in proper development of tube well is also responsible for low yield.

3.10.3 *Well failure due to incrustation* – The reduction in pressure around the screen causes some of the salts dissolved in water to precipitate on the screen openings (regardless of its material type). Incrustation because of carbonates is considered as *'soft'* and can be removed easily. Generally, $CaCO_3$ is a chief incrusting agent of soft incrustation. Along with $CaCO_3$, $Fe(OH)_2$ and $Fe(OH)_3$ are responsible for more than 90% of mineral incrustation problems (Mogg, 1972). The precipitated matter can be dissolved easily by hydrochloric acid (HCl) or sulfamic acid (H_3NSO_3). To avoid corrosion of screen material, inhibitors are added to acids to slow this tendency. Precipitation of water-soluble sulphates is considered under the *'hard'* category and is not due to pressure reduction. Water temperature is responsible for this precipitation as at higher temperature, solubility of calcium sulphate ($CaSO_4$) decreases.

Iron bacteria are mainly responsible for the bacterial plugging of the screens. The bacteria belong to this generation are Genus gallionella, Crenothrix, and Leptothrix. These bacteria oxidize the dissolved iron and manganese, which causes it to precipitate. The iron bacteria and plugging caused by it is most prevalent in the following situations (Mogg, 1972):

- Under aerobic conditions (generally dominant in shallow wells).

- Groundwater having temperature less than 18°C.
- Groundwater containing high concentrations of iron and manganese (more than 1 ppm).
- Groundwater having total dissolved solids (TDS) less than 1000 ppm.

For killing iron bacteria, at least 18 h contact period with the chlorine having concentration of 300 ppm is required (Grainage and Lund, 1969). Chlorine dioxide (ClO_2) and hydroxyacetic acid ($C_2H_2O_3$) are also used for killing iron bacteria. Chlorine dioxide is a synthetic, green-yellowish gas with chlorine like irritating odor. It is highly soluble in water (specifically in cold water) and is a strong oxidizer. Hydroxyacetic acid is a colorless, odorless, hygroscopic crystalline solid, which is highly soluble in water.

3.10.4 *Corrosion* – It is the eating away or destruction of metal by mean of a chemical reaction between the metal and the environment. Corrosion of the well screen may cause failure of tube well in the following ways:

- Slot size of the screen gets enlarges, which promotes sand pumping.
- Strength of the screen is reduced by eating away of metal which may lead to collapse of well screen or casing.
- There are chances that the corroded material may be re-deposited in the screen openings. As a result, there is a reduction in the tube well yield.

4

Water Lifting Devices

4.1 Introduction

Lift irrigation involves the lifting of water from its source to the application level. The selection of a lifting device in a particular situation depends upon the following criteria

- Available water source
- The depth of water level from the point of application
- Quantity of water required
- Type of power available
- Economic status of the farmer

The indigenous water lifts came in to existence the time immemorial. However, the present day indigenous water lifts are much-refined versions of the primitive lifts. The users all over the world have either invented or modified the design of various water lifts according to their needs. For operating most of the indigenous water lifts, either manual or animal power is required. On the other hand, with the advent of highly efficient electric motor/engine operated modern centrifugal pumps, a revolution occurred in the lift irrigation schemes. Wind and water-powered lifts are location specific devices. Because of lesser capacity and close running parts, which may wear out with the presence of foreign particles in water, positive displacement pumps are not popular in lift irrigation schemes.

4.2 Classification of irrigation water lifts

4.2.1 *Indigenous water lifts* – The classification of commonly used irrigation water lifts is presented in the schematic chart (Figure 4.1). Before the advent of diesel engine and electric motor operated pump sets, several types of indigenous water lifts were prevalent in different parts of India. Depending upon the depth of water from the level of application, these indigenous techniques are grouped in low, medium, and high headwater lifts (Figure 4.2).

4.2.2 *Wind-powered water lift* – The discharge from a pump (reciprocating) operated by wind power is intermittent and seasonal as its capacity depends on the wind velocity and frequency. For the satisfactory operation of windmill, wind velocity more than 6.5 km/h is necessary. However, high wind velocity areas in India are limited to coastal regions and that also during the summer months when monsoon is active.

4.2.3 *Water-powered water lift (Hydraulic ram)* – Hydraulic ram is used to raise a part of large quantity of water available at certain height to a greater height (Figure 4.3). It is also known as *impulse pump* and the impulse is developed by the sudden obstruction caused (by valves) to moving column of water. The magnitude of the impulse can be increased by increasing the length of supply pipe. If, Q is the volume of water escaping from the waste valve, q is delivering by ram, H is the drive head, and h_d is the delivery head, then the efficiency of the hydraulic ram can be evaluated by D'Aubuission's ratio as

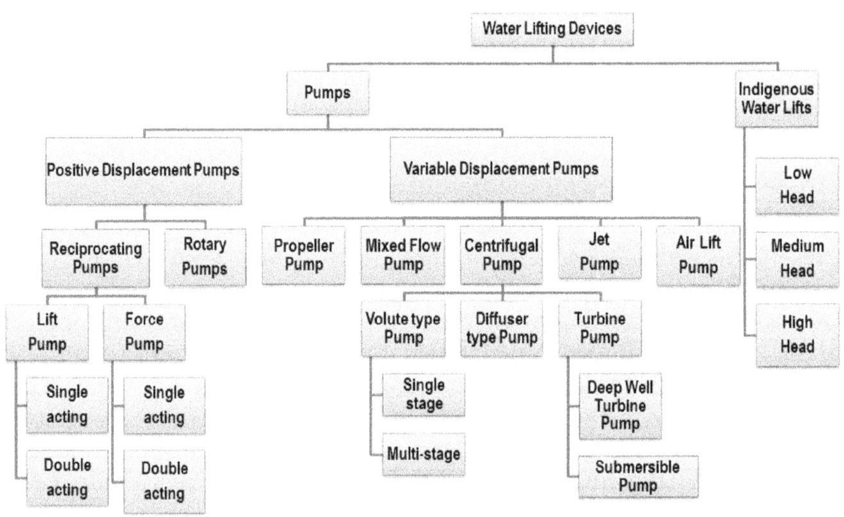

Figure 4.1: Classification of water lifting devices.

Low Head Water Lifts

Swing Basket: Common in rice growing belts of eastern and southern India. Two men can lift water at a rate of around 14,000 to 19,000 ℓ/h from a depth of 0.9 to 1.2 m.

Don: Common in eastern regions of India. One man can lift the water at a rate 9,000 to 13,000 ℓ/h from a depth of 0.5 to 1.0 m.

Archemedian Screw: Most common in deltaic regions of Andhra Pradesh. One man can lift water at a rate of around 14,000 to 19,000 ℓ/h from a depth of 1.0 to 1.2 m.

Water Wheel: It was the common practice in the canal irrigated areas of northern India (now abandoned). One-pair of bullock and one operator can lift water at a rate of 40,000 to 60,000 ℓ/h from a depth of 1.0 to 1.2 m.

Medium Head Water Lifts

Persian Wheel (*Rahat*): Most common practice in the northern India before the adaption of diesel/electric operated pumps. Depending upon the availability of animal power (i.e. bullocks, buffaloes, and camel) and one operator, water at a rate of 14,000 to 18,000 ℓ/h can be lifted from a depth of 5.0 to 10.0 m.

Chain Pump: Was in practice in some parts of U.P. A pair of bullock and one operator can lift water at a rate of 15,000 to 20,000 ℓ/h from a depth of 3.0 to 6.0 m.

Self-emptying type rope-and-bucket lift: Common in southern and Deccan regions of India. One-pair of bullock and one operator can lift water at a rate of 10,000 to 15,000 ℓ/h from a depth of 4.0 to 6.0 m.

Circular two-bucket lift: Common in some parts of Tamil Nadu. An operator with a single bullock can lift water at a rate of 12,000 to 14,000 ℓ/h from a depth of 4.0 to 5.0 m.

Counterpoise bucket lift (*Dhenkli* or *Piccottah*): One-man can operate this system and is common in Bihar and southern India. Around 8000 to 11,000 ℓ of water can be lifted in a hour from a depth of 1.2 to 4.0 m.

High Head Water Lifts

Rope and Bucket lift (*Charas*): Common in areas having deep water table (such as in areas of Rajasthan and Maharashtra). Maximum power was required in this lift. Two pairs of bullocks and three men can lift water at a rate of 6,000 to 10,000 ℓ/h from a depth of 10.0 to 30.0 m.

Figure 4.2: Classification of indigenous water lifts with reference to head.

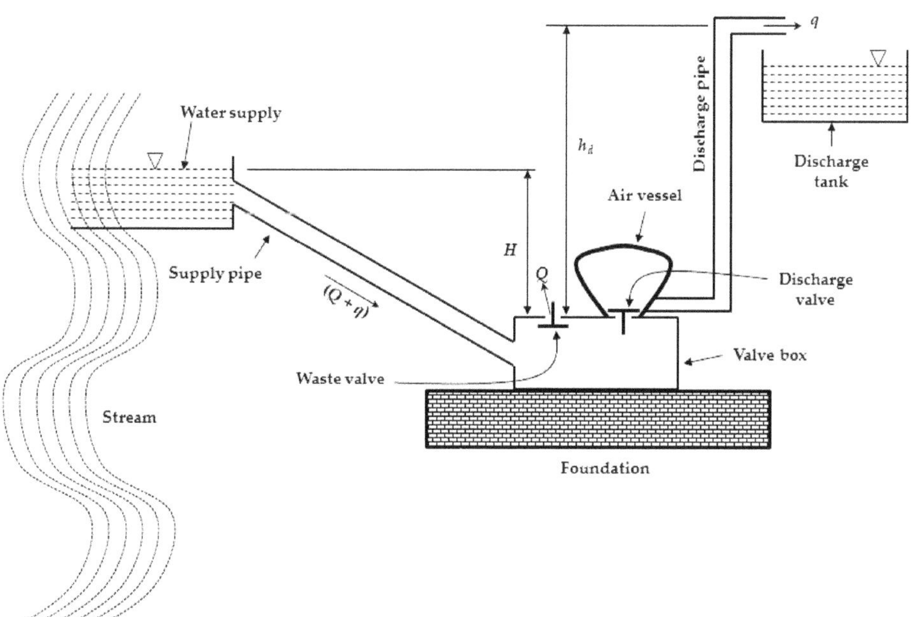

Figure 4.3: Schematic sketch of a hydraulic ram.

$$\frac{qh_d}{(Q+q)H} \tag{4.1}$$

4.3 Positive displacement pumps

The distinguished feature of *positive displacement pump* is the discharge of same volume of water irrespective of the head against which they operate. Because of small discharging capacity, these pumps are not used for irrigation or drainage purposes. Reciprocating pump (also known as *piston* or *displacement pump*) is an example of positive displacement pump (Figure 4.4). The size and stroke length of a piston or plunger decides the capacity of pump. To prevent leakage between piston and the cylindrical chamber in which it is contained, *cup-leather packing* is used.

Theoretically, pump cannot lift water from a suction depth of more than 10.335 m (or 1 atmosphere). Moreover, this theoretical limit is valid at sea level only. In actual field applications, elevation from the sea level, friction, and other losses decrease this limit to about 6 – 7 m.

4.3.1 *Animal powered reciprocating pump* – It can be used to lift water when the depth of pumping water level is not more than 6 m. As per Khepar et al. (1975), this pump has a capacity to discharge 25,200 l/h against a head of 4 m.

4.3.2 *Single and double acting reciprocating pump* – This pump has only one discharge stroke for every two strokes of the piston and therefore, water is delivered in every alternate stroke (intermittent flow). For ensuring nearly continuous flow, an air vessel at or near the discharge valve can be fixed. The construction arrangement of double acting reciprocating pump is shown in Figure 4.5. Considering that the pump cylinder and all the connecting pipes are filled with water, the volume of water delivered in each stroke will be equal to the volume swept by plunger in one

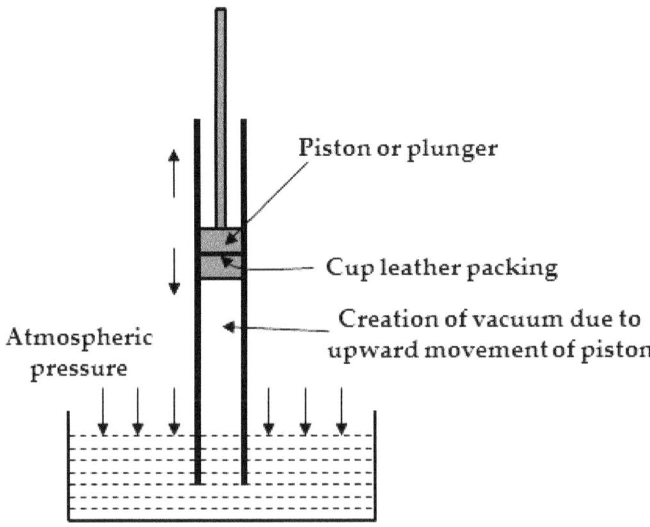

Figure 4.4: Schematic sketch showing the operation of piston pump.

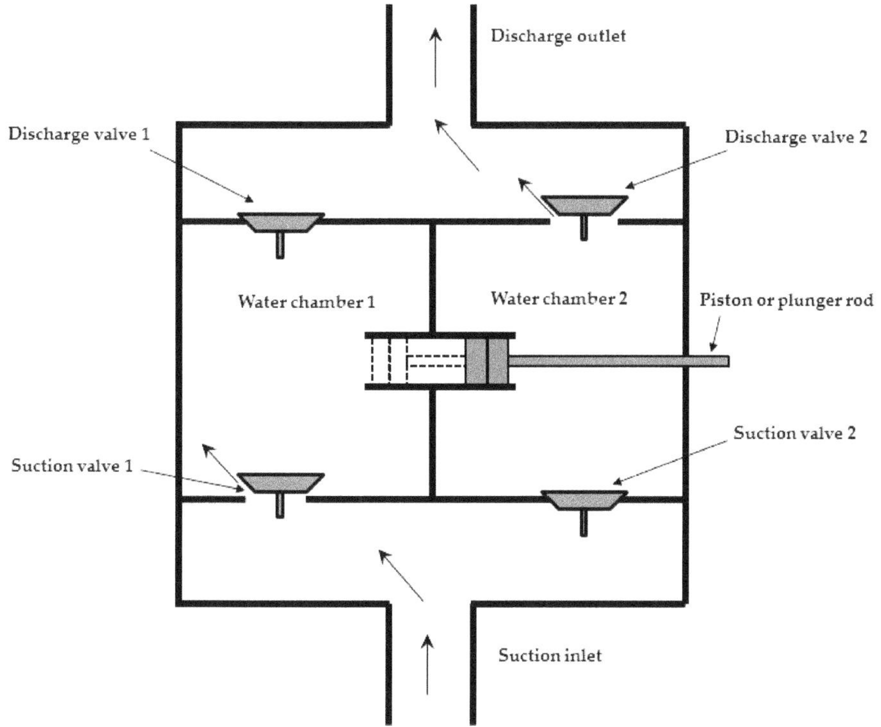

Figure 4.5: Schematic sketch showing the operation of double acting piston pump.

stroke. Neglecting the volume of plunger rod and friction losses, the force required to work a reciprocating pump can be calculated by the formula as

$$P = wal \tag{4.2}$$

In Eq. (4.2), w is the specific weight of water (i.e. 1000 kg/m^3); a is area of cylinder (m^2); and l is length of plunger stroke (m).

4.3.3 *Rotary pump* – In *rotary pump*, closely meshed gears or cams are housed in a chamber with minimum clearance. Regardless of the pressure against which pump is working, discharge delivery is constant through it.

4.4 Variable displacement pumps

In *variable displacement pumps*, the discharge rate is inversely related to pressure head.

4.4.1 *Specific speed* (n_s) – It is a term used to classify the impellers based on their performance and proportions regardless of their actual size or the speed at which they operate. It is defined as the speed in rpm at which an impeller would operate if reduced proportionately in size so as to deliver a rated capacity of 1 m^3/s against a total head of 1 m (Church, 1944). Mathematically, it can be written as

$$n_s = \frac{nQ^{\frac{1}{2}}}{(H\,/\,\text{Stages})^{\frac{3}{4}}} \tag{4.3}$$

In Eq. (4.3), n is the pump speed (rpm), Q is the pump discharge (m^3/s), and H is the total head (m). The specific speed for single suction impellers ranges from 10 to 300. While comparing the specific speed of a double suction impeller with the single suction impeller, the capacity of the former is divided by two.

4.5 Definitions

Pump capacity – It is the ability of a pump to discharge per unit time and is usually measured in the units of l/s, l/h, or m^3/d.

Suction lift – When the level of water is below the centerline of pump, *suction lift* exists (Figure 4.6). It is a qualitative measure.

Static suction lift – It is the vertical distance from the centerline of pump to the level of water (below centerline of pump) to be pumped (Figure 4.6). It is a quantitative measure.

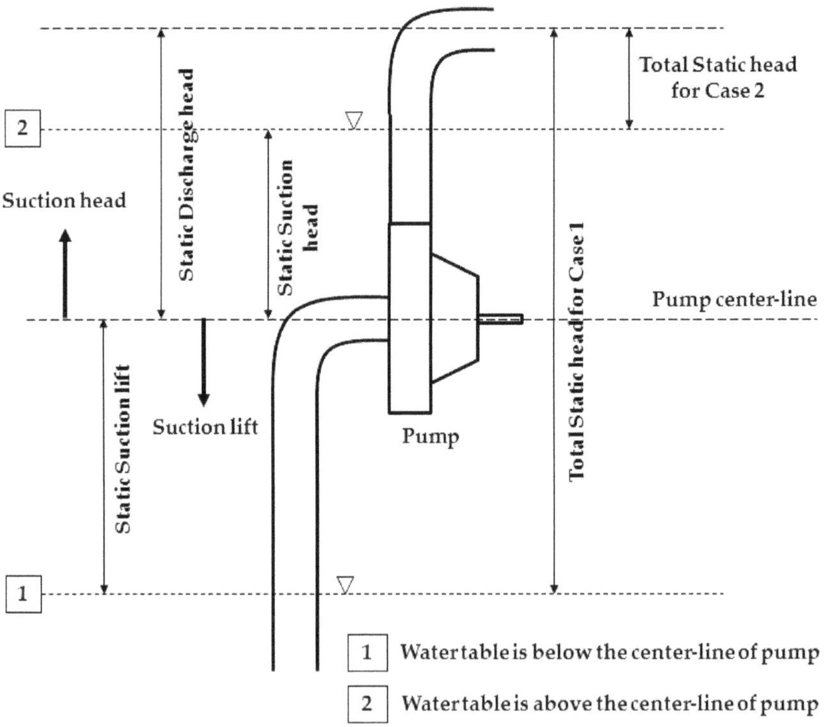

Figure 4.6: Schematic sketch depicting the pump related definitions.

Total suction lift – When the entrance losses and frictional losses through pipes and fittings (in terms of head) are added to the *static suction lift*, it becomes total suction lift (Figure 4.6).

Suction head – When the level of water is above the centerline of pump, suction head exists (Figure 4.6). It is also a qualitative measure.

Static suction head – It is the vertical distance from the centerline of pump to the level of water (above centerline of pump) to be pumped (Figure 4.6). It is a quantitative measure.

Total suction head – When the entrance, any pressure head on the suction line, and frictional losses through pipes and fittings (in terms of head) are subtracted from the *static suction head*, it becomes total suction head (Figure 4.6).

Static discharge head – It is the vertical distance of the free discharge level from the pump centerline (Figure 4.6).

Friction head – Head required to overcome the friction caused to flow through pipes and fittings (expressed in meters of water) is known as friction head.

Pressure head – It is the pressure (when expressed in meters of water) developed by the pump. It is the energy contained in the water because of its pressure and can be calculated as

$$h_p = \frac{P}{w} \tag{4.4}$$

Velocity head – It is the measure of the kinetic energy contained in a unit weight of water due to its velocity and can be calculated as

$$h_v = \frac{v^2}{2g} \tag{4.5}$$

Total discharge head – When the exit losses, frictional losses through pipes and fittings (in terms of head), velocity and pressure head (at the point of discharge) are added to the *static discharge head*, it becomes total discharge head (Figure 4.6).

Total static head – It is the vertical distance from the suction water level to discharge water level. In Figure 4.6, total static head for the two cases has been shown.

Net positive suction head (*NPSH*) – When the absolute pressure of the water reaches its vapor pressure, cavitation occurs. *NPSH* is the total head at the pump centerline. For its determination, following factors are required

- Absolute pressure on the surface of water body (H_p). For unconfined aquifers and open water bodies, it is atmospheric pressure corresponding to the altitude.
- Vertical distance from the pump centerline to the pumping water level (H_z). If water level is below the pump centerline, it is considered negative and vice-versa.
- Vapor pressure of water (in terms of head) at the existing temperature (H_{vp}).

- Friction losses in the screen, pipe, fittings, bends on the suction line (H_f).

Combining all the above-mentioned terms, the equation for *NPSH* can be written as

$$NPSH = H_p \pm H_z - H_{vp} - H_f \tag{4.6}$$

For every 300 m increase in altitude, atmospheric pressure (10.35 m at sea level) reduced by 0.36 m.

Water horsepower (*WHP*) – The *WHP* is the product of the delivered flow in liters per second (l/s) times the total head in meter (m) developed by the pump. It is the theoretical horsepower and can be written as

$$WHP = \frac{Pumping\ rate\ (l/s) \times Total\ head\ (m)}{75} \tag{4.7}$$

Equation (4.7) in different units can be written as

$$WHP = \frac{Dischare\ \left(m^3/h\right) \times Total\ head\ (m)}{273} \tag{4.8}$$

Shaft horsepower (*SHP*) – It is the power required at the pump shaft which includes *WHP*, the impeller friction, circulation losses, stuffing box and bearing friction losses. Mathematically, it can be written as

$$SHP = \frac{WHP}{Pump\ efficiency\ (\eta_p)} \tag{4.9}$$

Break horsepower (*BHP*) – It is the actual power required to drive the pump. Further, it depends upon the mode of drive from the prime mover (engine or electric motor). For uni-built and directly coupled pumps, *BHP* will be equal to *SHP*. For belt drives, *BHP* can be calculated as

$$BHP = \frac{SHP}{Drive\ efficiency\ (\eta_d)} \tag{4.10}$$

Incorporating Eq. (4.9) in Eq. (4.10), the formula modifies to

$$BHP = \frac{WHP}{Pump\ efficiency\ (\eta_p) \times Drive\ efficiency\ (\eta_d)} \tag{4.11}$$

For evaluating horsepower input to electric motor, motor efficiency also comes into picture and the Eq. (4.11) changes to

$$HP\ input\ to\ electric\ motor = \frac{WHP}{Pump\ efficiency\ (\eta_p) \times Drive\ efficiency\ (\eta_d) \times Motor\ efficiency\ (\eta_m)} \quad (4.12)$$

The power evaluated from Eq. (4.12) can be converted to kilowatt by multiplying it by 0.746.

4.6 Centrifugal pump

A centrifugal pump consists essentially of one or more impellers equipped with curved vanes (or blades), mounted on a rotating shaft, and enclosed by a casing (Church, 1944). In volute type centrifugal pump, the water enters the casing after leaving the impeller (Figure 4.7a). The cross-sectional area of the volute casing increases towards the discharge opening and its shape is usually in the form of volute curve (Figure 4.7b). In diffuser type centrifugal pump, a series of guide vanes surrounds the impeller (Figure 4.7c).

The high velocity is imparted to fluid by the impeller. As the high velocity fluid enters the casing, its velocity is reduced and converted into pressure energy. In case of volute centrifugal pump, conversion of velocity energy to pressure energy is accomplished in volute casing; whereas, in case diffuser type centrifugal pump the major conversion is taken place in the diffusers.

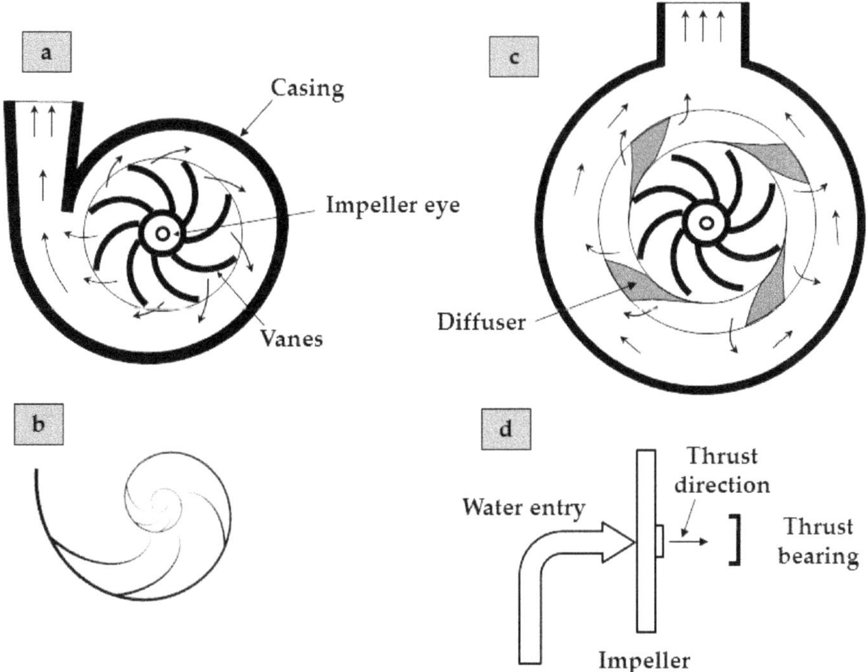

Figure 4.7: Schematic sketches of (a) volute centrifugal pump, (b) volute curve, (c) diffuser/turbine centrifugal pump, and (d) role of thrust bearing.

The quantity of water that a pump can deliver against a given head can be increased by enlarging the width of impeller (i.e. width of vanes) and its eye diameter.

For delivering water against high head conditions, an impeller with more number of vanes is used. However, for high pumping rate, impeller with lesser number of vanes is used. In general, the number of vanes on the impeller ranges from 2 to 12.

To prevent leakage of water from the casing through the annular space where shaft passes through it, a seal or a *stuffing box* is provided. The materials like organic fiber and asbestos are used for manufacturing packing ring.

The thrust developed by the entry of water from one side of the impeller (Figure 4.7d) must be overcome by the hydraulic or mechanical means. The use of thrust bearing is a mechanical mean to bear thrust (Figure 4.7d). By providing suction at both ends of the impeller, water thrust can be countered hydraulically (Figure 4.8a). In case of multi-stage pumps, impellers can be mounted on shaft back to back (Figure 4.8b).

For reducing the leakage of water through the packing boxes where the shaft enters the casing, *labyrinth rings* are used. These are made of thin strips of stainless steel, lead, or bronze and may be of either staggered or straight-through type (Figure 4.9a-b).

4.6.1 *Characteristic curves* – The curves that show the inter-relations between capacity, head, power, and efficiency of a pump are known as characteristics curves (Figure 4.10). The curves vary with the speed of the pump. These are also known

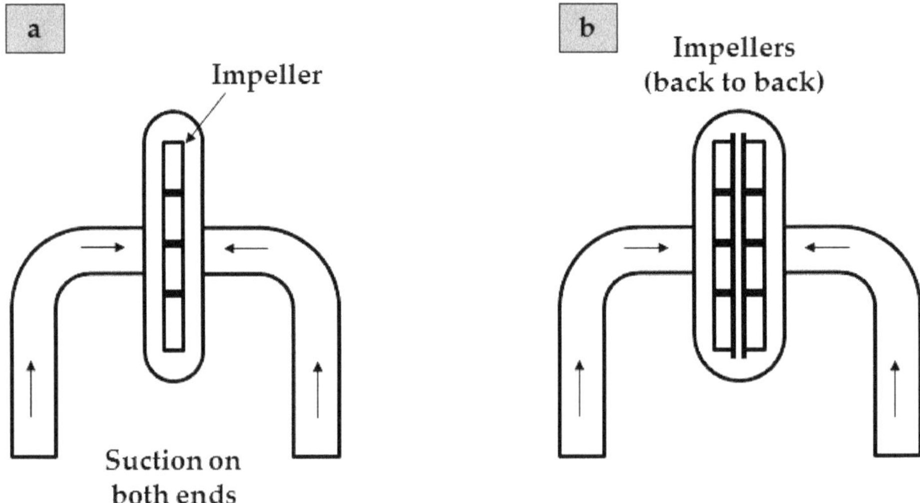

Figure 4.8: Schematic sketches showing centrifugal pump with (a) suction at two ends, and (b) back to back impellers for balancing the thrust hydraulically.

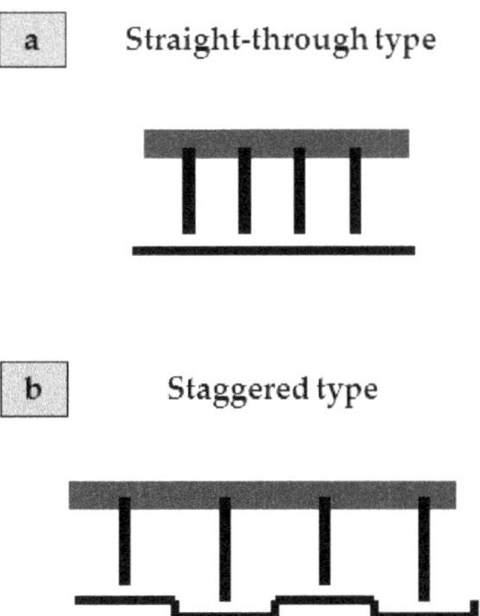

Figure 4.9: Schematic sketches showing the labyrinth ring of (a) straight-through type, and (b) staggered type.

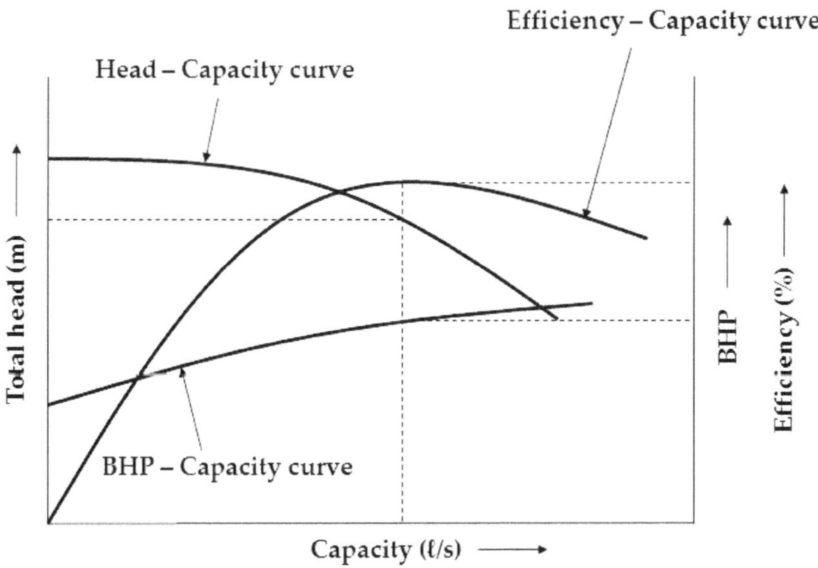

Figure 4.10: Graph showing the characteristic curves.

as *performance curves*. Based on characteristic curves, one can select the pump for particular conditions. During the pump test, about 6 to 12 readings are taken and are plotted. From Figure 4.10, it is clear that at zero discharge, efficiency of the pump is also zero. Moreover, with increase in head, discharge of the pump decreases. With the increase in the capacity, BHP of the pump increases. The method of reading the graph has been shown by dotted lines on it.

4.6.2 *Composite characteristic curves* – When head-capacity curves are drawn for one impeller diameter at different speeds or at one speed with different impeller diameters, these are known as composite characteristic curves. On these curves, points of equal efficiency are joined to form iso-efficiency lines.

4.6.3 *Effect on pump characteristics with change in impeller speed and diameter* – With change in the speed of impeller, the capacity, head, and power of the pump changes in the following way.

- Capacity of the pump varies directly with the impeller speed.
- Head varies with the square of the impeller speed.
- BHP varies with the cube of the impeller speed.

All the relations are written in the Figure 4.11. Change in the impeller diameter affects the pump performance in the similar way as change in the speed.

4.6.4 *Priming* – Unlike positive displacement pumps, centrifugal pumps cannot compress the fluid. Hence, to initiate pumping, priming of centrifugal pumps is required. In priming operation, pumps are filled with water up to the top of the pump casing. For priming the centrifugal pump or to maintain it in the primed condition, following devices are used

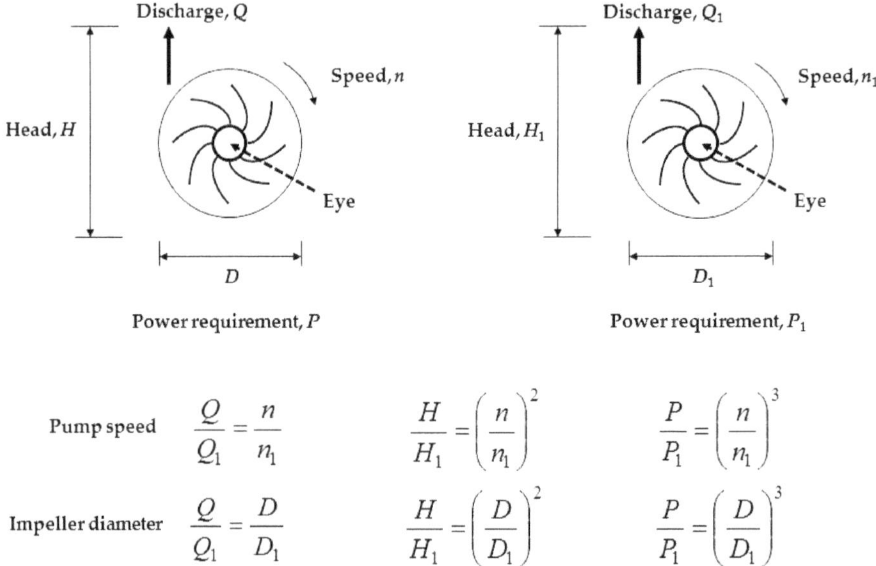

Figure 4.11: Schematic sketches and formulae showing the effect of impeller speed and diameter on the pump performance.

Reflex valve Inside view of reflex valve

Figure 4.12: Photographs of the reflux valve (outside and inside view).

- Reflux or foot valve (Figure 4.12)
- Auxiliary piston pump
- Connection to water source at high head
- Self-priming construction in pumps

4.6.5 *Classification of centrifugal pumps* – Based upon the different working conditions, centrifugal pumps are manufactured with various specifications (Table 4.1)

4.7 Points to remember while installing centrifugal pump

For prolonged and trouble-free service, centrifugal pump should be installed keeping the following points in view.

- As far as possible, centrifugal pump must be installed close to the water surface. By doing so, the length of suction pipe will be less and maximum capacity will be achieved.
- Installation location should be easily accessible, clean, dry, and properly ventilated.
- The foundation of the pump should be enough rigid so that it could absorb all vibrations.

4.7.1 *Pump foundation* – The foundation should be rigid enough to absorb all vibrations. It may be of

- Cement concrete base
- Steel angles
- Wooden beams

Table 4.1: Classification of centrifugal pumps.

1.	**Energy conversion in the section**	
	Volute – The volute casing is proportioned to reduce the velocity of water from the start of the volute section to end. With the increase in area, velocity head changed to pressure head gradually.	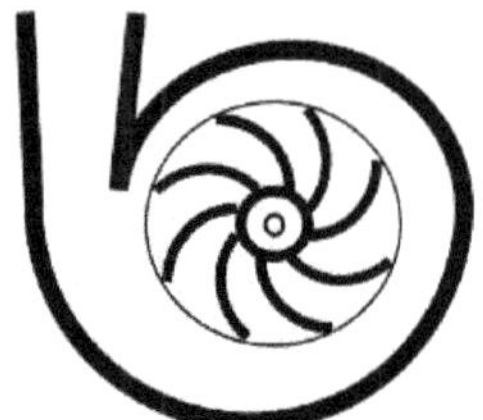
	Diffuser – The openings of the diffuser vanes enlarge gradually to their outer diameter, where the major conversion of the velocity into pressure head occurs.	
2.	**Stages**	
	Single – A pump with single impeller is also known as single-stage pump.	
	Multi-stage – When two or more impellers are fixed in series on a single shaft, it is known as multi-stage pump. In multi-stage pump, water from the discharge of first impeller enters the succeeding one and so on. The typical characteristics of a multi-stage pump are as follows • The head and power requirement of a multi-stage pump can be evaluated by multiplying the number of stages (or impellers) with the head and power requirement of single stage (or impeller). • The pumping capacity and efficiency remains the same as for a single stage pump operating alone.	

Table 4.1: (Continued)

3.	**Type of impellers** – The efficiency and operating characteristics of centrifugal pump are greatly influenced by the design of impeller. The impellers may be classified as	
	Open impeller – The vanes are attached to the central hub. Convenient to pump water having small solids.	
	Semi-open/semi-enclosed impeller – The sidewall or shroud on one side only. Convenient to pump water having suspended sediments.	
	Enclosed impeller – The blades or vanes are enclosed between sidewalls or shrouds on both sides. Preferable to pump sediment-free water.	
	Non-clog impeller – It is specifically designed for handling the sewage, which generally contains solids, rags, and other materials.	
4.	**Suction inlets**	
	Single – In single suction pump, water enters the impeller from one side. (Generally used)	

(Continued)

Table 4.1: (Continued)

	Double – In double suction pump, water enters the impeller from both sides. (Difficult to manufacture)	
5.	**Casing split**	
	Vertically split - Casing plane is perpendicular to the pump shaft.	
	Horizontally split – The horizontally split casing is also known as *axially split*.	
6.	**Axis of rotation**	
	Horizontal – Impeller remains in vertical position in this type of pump. It is the most commonly used pump in irrigation.	
	Vertical – In this case, the impeller remains in horizontal position and its axis is vertical.	
7.	**Pump driving method**	
	Flexible belt joints (specifically for engine operated pumps) – In this case, flexible belts are used to connect engine with pump. (Misalign joint can also work in such type of arrangement).	

Table 4.1: (Continued)

	Directly coupled motor – Flexible couplings are used to connect motor with pump. (Minor misalignment is permissible in this case).	
	Close-coupled/uni-built pump and motor – In this type of arrangement, pump and motor are built with a common shaft and bearings.	
	Belt – Usually adapted when the power source is located away from the pump or when we operate multiple machines with single power unit. The pump can be engaged and disengaged by shifting the belt on the fast or loose pulley, respectively.	

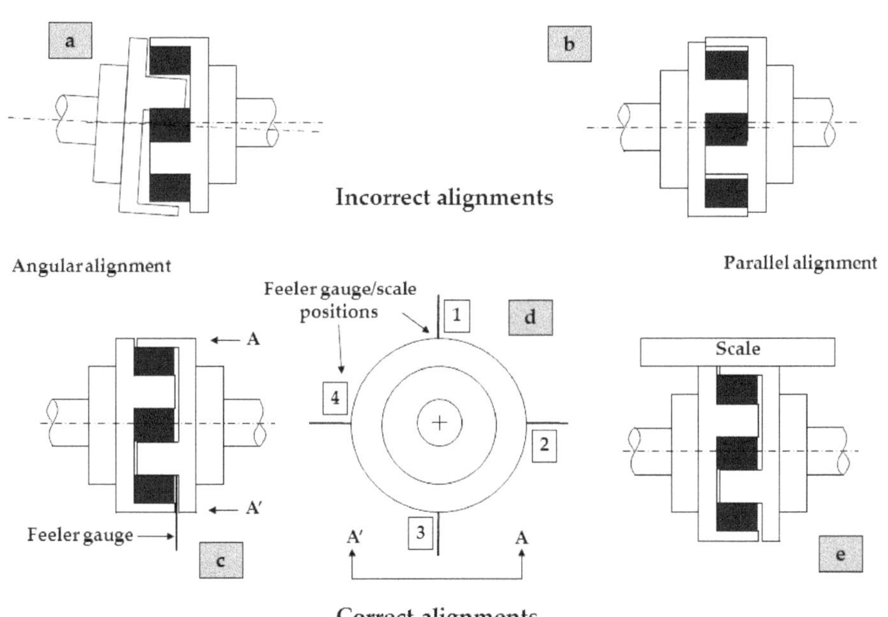

Figure 4.13: Schematic sketches showing the methods of judging the angular and parallel alignment.

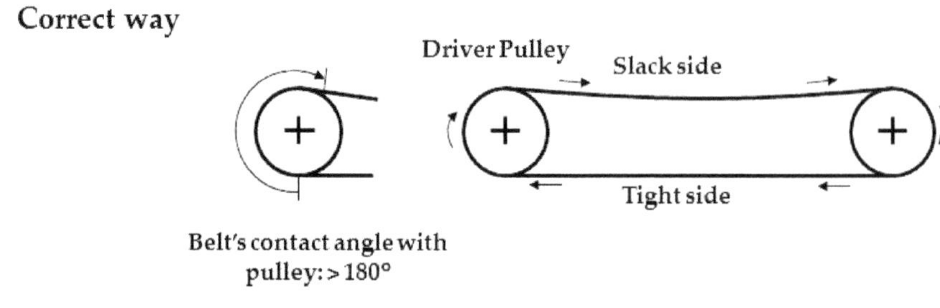

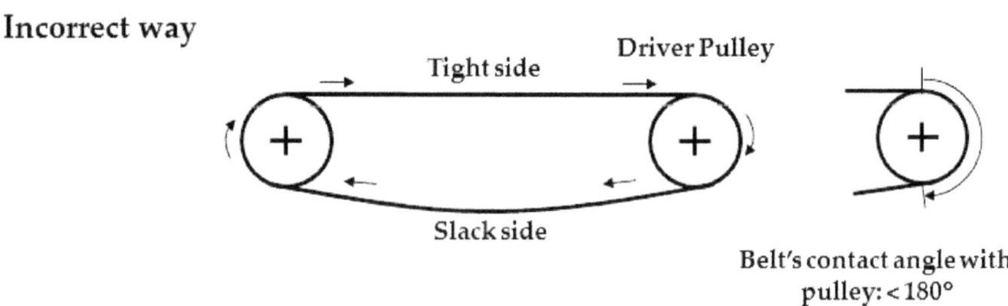

Figure 4.14: Schematic sketches showing the correct and incorrect way of operating pulley drives.

4.7.2 *Alignment* – The alignment between pump and driver should be checked from the angular and parallel alignment point of view (Figure 4.13). For checking the *angular alignment*, the feeler gauge is inserted between the coupling halves at four positions, which are at 90° interval around the coupling (positions 1 to 4 are shown in Figure 4.13c). For checking the *parallel alignment*, a straight edged scale is positioned across the coupling halves at four positions as shown in Figure 4.13d. For belt driven pumping units, the shafts of the pump and the driver should be parallel. In case of V-belt and flat belt drives, the tight side of the belt should be at the bottom (Figure 4.14). By doing so, the slackened side of the belt (due to its own weight) remains in touch with pulley for longer period than the other option.

It is advisable to avoid using the flat belts for vertical derives. However, in industries where single power unit is used for multiple purposes, the adaption of vertical derives is a common practice. Practically, in case of flat belts an angle of more than 45° between the line joining the shafts and horizontal should be avoided (Figure 4.15a). Moreover, the speed of the flat belt is limited to about 25 m/s. Diameter of one pulley should not exceed the diameter of the other pulley by 5 times (Figure 4.15b). For removing the slack of belt, it should not be tightened excessively; otherwise, the excessive tension will overload the bearings and loss in energy.

4.7.3 *Suction piping* – While installing suction pipe, the following points must be remembered.

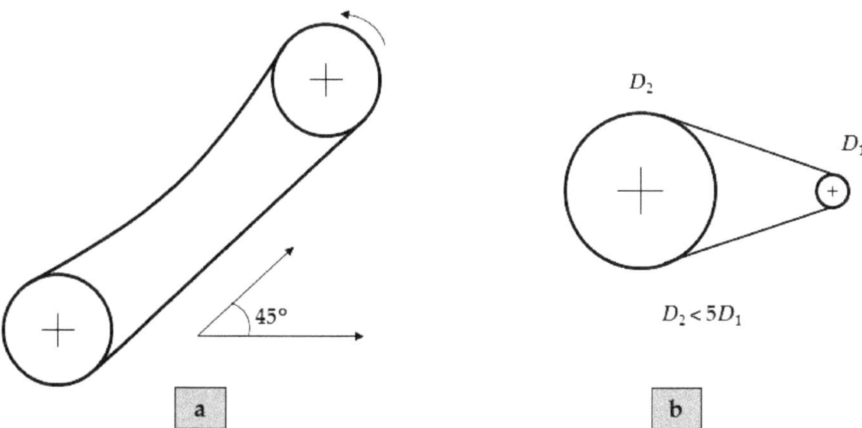

Figure 4.15: Schematic sketches showing pulley drives having (a) angle of 45°, and (b) diameters in 1:5 ratio.

- Suction pipe should have minimum number of fittings to avoid unnecessarily increase in frictional losses.
- Joints should be airtight.
- Sharp angle bends should not be used.
- The size of the suction pipe should be equal or more than the size of pump inlet. Its size is so selected that the velocity of flow through pipe should not exceed 3 m/s. The flow velocity more than this limit may lead to the cavitation.

4.7.4 *Foot valve* – It is an essential component of the suction piping while pumping from open water bodies (such as pond, open well, canal, river). Purpose is to keep the pump in primed condition. The following points should be kept in mind while installing it.

- Foot valve should be submerged to at least 60 cm below the pumping water level.
- The total perforated area in the foot valve should be 3 to 4 times the cross-section area of the suction pipe.
- Foot valve should not be kept on the bed of the water body to prevent its chocking up by mud or other material.

4.7.5 *Reflux valve* – For the screened wells (i.e. installed in the boreholes), reflux valve is used to keep the pump in primed condition. Generally, the reflux valve is installed on the suction pipe near to pump (Figure 4.12). During pumping operation, its valve opens at 90° angle.

4.7.6 *Tractor operated pumps* – Tractors can be used to operate centrifugal pumps. The power for running pump is taken from tractor power take off (PTO). However, there is need to install a speed increasing mechanism as the speed of tractor PTO is around 540 ± 10 rpm; whereas, pump speed usually exceeds 1450 rpm. The farmers in certain pockets of Punjab are using the tractor power to generate electricity (by attaching alternator to PTO), which in turn is used to operate the electric motor driven pump.

4.7.7 *Role of A.C. motor starter* – For 3-phase motors, starters are required. If we try to start the motor on full line voltage, it will draw about 4 times its normal running current that may damage the motor. For the purpose, Star-Delta starter is used, which work on the reduced voltage starting principle. While starting, *star* connections are used and this reduces the line voltage to $\frac{1}{\sqrt{3}}$ times the phase voltage. The phase current is also reduced in the same proportion. As the motor accelerates, the connections are switched over to *delta*. To protect the motor from single phasing and overload, starters are also equipped with safety mechanisms (which may be based upon thermal or magnetic principles). At about 10% above the maximum motor current, the overload release mechanism trips off the motor.

4.8 Cavitation

If the water pressure at any point inside a pump drops below the vapour pressure corresponding to the temperature of water, the water will vaporize and form cavities of vapour. These vapour bubbles are carried along with the stream until a region of higher pressure is reached when they collapse or implode with a tremendous shock on the adjacent walls (Church, 1944). Basic causes of occurrence of cavitation phenomenon are due to improper design, installation, and operation of pump. The cavitation can be prevented by adapting the following practices

- Pump should be operated under the conditions for which it is designed.
- Lowering of water level with respect to pump location increases its suction lift that may cause cavitation phenomenon. Continuous monitoring of water level is necessary for proper functioning of the pump.
- Pump should not be run at a speed more than its rated value.

4.9 Common troubles in centrifugal pumps

Sometimes, centrifugal pump fails to lift water or develop rated pressure. Lack of proper maintenance is the major reason for this. However, other factors (sometimes) are responsible for these problems. In Table 4.2, such troubles and their causes are listed for ready reference of users.

4.10 Water hammer

When water flowing through a pipe undergoes a sudden change in the velocity, *hydraulic shock* occurs. During switching off the motor, upward moving column of water through the vertical pipe suddenly stops and reverses its direction due to gravity. This falling column of water creates dynamic pressure wave, which may produce substantial impact on the impeller and pump casing. In long pipe, such impact sometimes reaches to destructive magnitude. Water hammer occurs during

Table 4.2: Common troubles of centrifugal pump, their causes, and the most probable reasons.

S. No.	Trouble	General causes	Reasons
1	Pump is not delivering water	Net positive suction head (NPSH) is beyond the permissible limit.	Lowering of water table.
		Pump is not primed.	Draining of water from the pump chamber due to improper seal of foot/ reflux valve. Leakage of water from stuffing box, and joints on suction side.
		Air is entrapped in suction line or pump casing.	Joints are not airtight.
		Severe leakage from stuffing box and joints on suction side, which are responsible for non-maintenance of primed conditions in the pump.	Non-usage of pump for longer duration that makes the stuffing box seals rigid enough to prevent leakage. Sealants are not properly inserted in the joints. Use of rubbers (local method) instead of clamps while joining flexible pipe to the pump inlet.
		Discharge head is beyond the limit of pump.	Incorrect measurement of discharge head (specifically in hilly regions).
		Foot valve or reflux valve is jammed and is not opening.	Junk on the hinge (of reflux/foot valve) make the valve inoperative.
		Pump speed too low.	Low-voltage due to some technical problem at the power distribution center. Diesel/petrol engine is not running properly. Low-intensity sun light (in case of solar operated pumps).
2	Discharge rate of the pump is low	Pump is running in the wrong direction.	Interchanging of electric phases by the electrician while making the connections.
		Discharge head nearly approaching shut-off head.	Incorrect measurement of discharge head.
		Wrong placement of foot valve and/or in-sufficient screened portion for water entry.	Foot valve is placed on the bed of water body. Screened portion is not according to the design.
		NPSH is approaching near the maximum permissible limit.	Declining water table is the prime factor responsible for this.
		Partial blocking of the suction pipe.	Bending of the flexible pipe. Entrance of foreign material in the suction pipe.
		Air leakage in the stuffing box and/or suction line.	Cracking of flexible pipe. Improper joints. Rigidness of the stuffing box seal.

(Continued)

Table 4.1: (Continued)

S. No.	Trouble	General causes	Reasons
3	Pump is unable to develop rated pressure	Rated head range of the selected pump is lower than the required. In other terms, it can be said that impeller diameter is small.	Wrong selection of pump.
		Pump is running in the wrong direction.	Interchanging of electric phases by the electrician while making the connections.
		Pump speed too low.	Low-voltage due to some technical problem at the power distribution center. Diesel/petrol engine is not running properly. Low-intensity sun light (in case of solar operated pumps).
4	Excessive requirement of power for running pump	Pump is running in the wrong direction.	Interchanging of electric phases by the electrician while making the connections.
		Misalignment of directly coupled or belt driven pumps.	Wrong installation of motor and/or pump.
		Shaft of the uni-built pump is bent.	Manufacturing defect.
		Seal in stuffing box is too tight.	Wrong selection of seal.
		Impeller is touching the casing	Improper placement of impeller or thrust bearing.
5	Pump is making excessive noise and is vibrating	NPSH is approaching near the maximum permissible limit.	Declining water table is the prime factor responsible for this.
		Misalignment of directly coupled or belt driven pumps.	Wrong installation of motor and/or pump.
		Foundation bolts are not tight.	Bolts may get loose with time.
		Foundation is not rigid (specifically for portable system).	Enough weight is not placed on the temporary foundation.
		Worn out of bearings and wearing rings.	The pump is not serviced from the long time.
		Lack of lubrication.	The pump is not under use/serviced from the long time.
		Impeller is not fixed rigidly with the shaft.	May have loosened with time.
6	Flow of the pump is intermittent (Specifically for submersible pumps)	Drawdown is reaching the maximum practical pumping level.	Lowering of water table. Hydraulic conductivity of the aquifer is not enough.

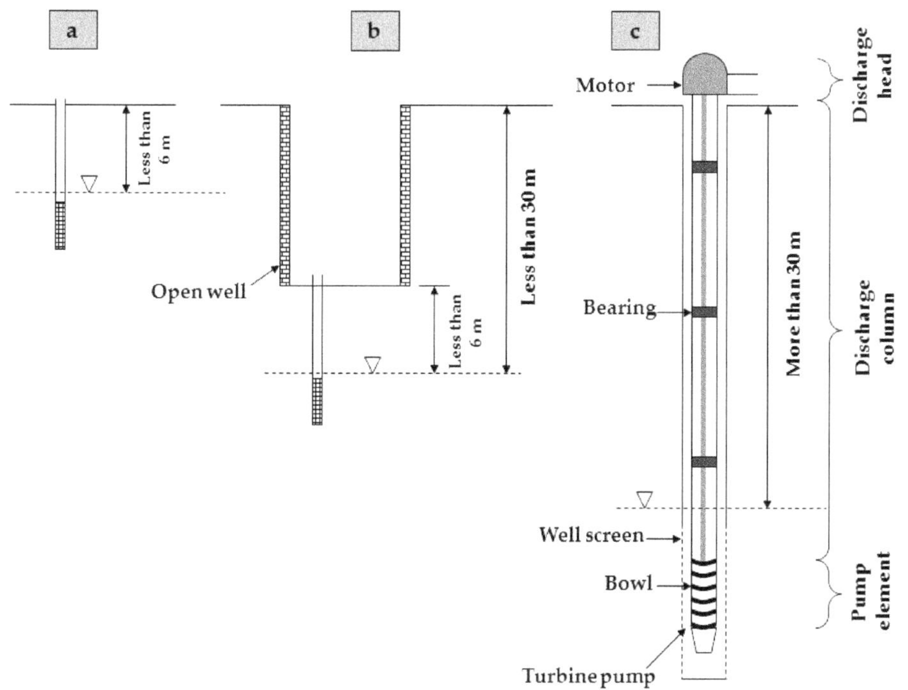

Figure 4.16: Schematic sketches showing the pumping conditions when water level is (a) less than 6 m, (b) between 6 to 30 m, and (c) more than 30 m from the ground surface.

(i) power failure, (ii) switching off of motor and, (iii) rapid closure of a valve in the connected pipe system. Department of Soil and Water Engineering, College of Agricultural Engineering and Technology, Punjab Agricultural University, Ludhiana, Punjab under the All India Coordinated Research Project on Groundwater Utilization (Anonymous, 2005) has developed a safety device to reduce the effect of water hammer.

4.11 Vertical turbine pump

It is commonly known as *deep well turbine pump*. It is well understood that suction lift for a centrifugal pump should not be more than 6-7 m (Figure 4.16a).

As the water level depth becomes more than 7 m, the installation position of the centrifugal pump should also be lowered to a depth so that the difference between water level and pump becomes less than 6 m. For installing the centrifugal pump below the ground surface, open wells up to required depth are dug out (Figure 4.16b). However, it is difficult to dig out open wells beyond certain depth (i.e. 15 – 20 m). For pumping water from the depths more than 30 m, a pump was required which could fit in the bore well and have enough capacity to pump against high head conditions. Therefore, a multi-stage pump was invented, which can be lowered in the borehole and worked under submerged conditions (Figure 4.16c).

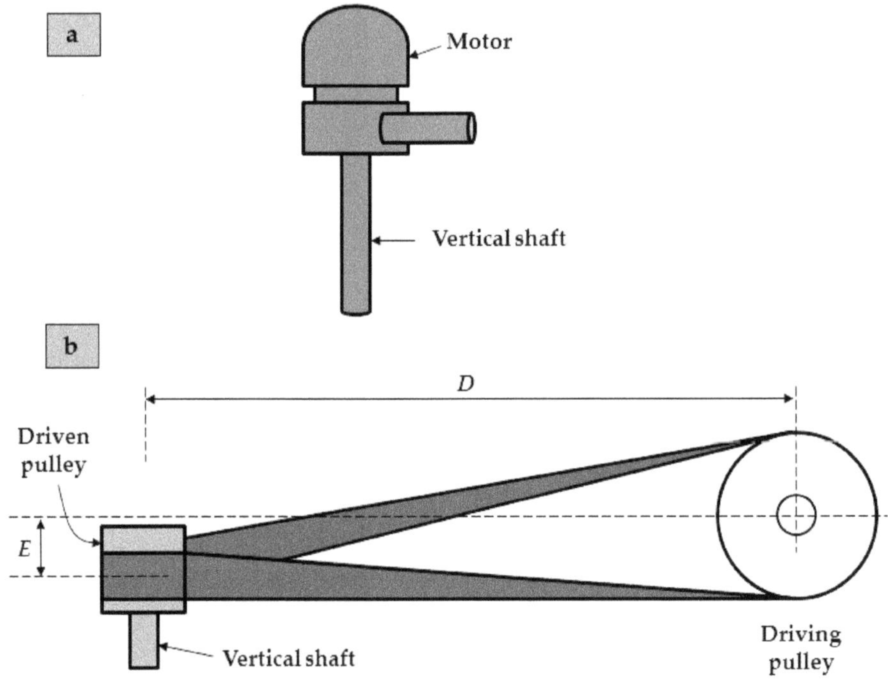

Figure 4.17: Schematic sketches showing the (a) directly coupled motor and (b) belt drive for vertical turbine pump.

The pump is operated from the ground surface with the aid of vertical axis. Vertical turbine pump can be divided into three parts, (i) *pump element*, (ii) *discharge column*, and (iii) *discharge head*. Pump element consists of multi-stage pump along with a small strainer at the bottom. Water from the underground formations is conveyed to the surface through discharge column. It also consists of vertical shaft, which is supported by bearings (Figure 4.16c). Discharge head assembly is like a base from which the discharge column and pump element suspends. It also incorporates the electric motor or engine operated pulley system. The characteristic curves for the vertical turbine pump are similar to the volute centrifugal pump (Figure 4.10). The driver of vertical turbine pump is located at the ground surface. Therefore, the possible drives may include belt drives, right angle gear drives, or directly coupled motor (Figure 4.17a) in vertical direction. Flat or V-belt drives can be used to drive the pump. V-belt drive is more efficient than the flat belt as in the latter case the slippage is more while transmitting power.

For rotating the shaft of vertical turbine pump, a quarter-twist arrangement for flat belt is required (Figure 4.17b). To allow for the natural tendency of belt sagging, the driving pulley is raised 6 cm for every 1 m distance between driving and driven pulleys. For example, if D = 2.5 m, then according to the recommendation, E will be equal to 15 cm. Regular inspection is the key to trouble free service of vertical

Table 4.3: Vertical turbine pump troubles and remedies.

S. No.	Troubles	General causes
1	Insufficient pressure	Pump speed too slow. Improper impeller trim. Impellers are loose. Cavitation. Worn-out of wearing rings. Discharge piping system is leaking. Leakage in column joints. Pump is running in the wrong direction.
2	Insufficient capacity of pump	Pump speed too slow. Improper impeller trim. Impellers are loose. Impeller or bowl partially plugged. Joints are leaking. Strainer partially clogged. Suction valve (if present) partially closed. Lowering of water level. Pump is running in the wrong direction. Total pumping head is too high.
3	Pump is not delivering water	Pump suction broken. Suction valve (if present) is closed. Impeller is plugged. Strainer is clogged. Pump is running in the wrong direction. Vertical shaft has been broken or unscrewed. Impellers are loose.
4	Pump is using too much power	Pump speed is too high. Improper impeller adjustment. Improper impeller trim. Pump out of alignment or shaft is bent. Packing rings are too tight. Impellers are damaged.
5	Pump is vibrating	Motor is not balanced. Motor bearings are not properly seated. Motor drive coupling out of balance. Misalignment of pump, casting, discharge head, discharge column, or bowls. Discharge head misaligned by improper mounting or pipe system. Shaft of the pump is bent. Bearings are worn out. Impellers are clogged or foreign material in pump. Impellers are not balanced.
6	Pump is making noise abnormally	Motor is making noise. Pump bearings are running dry. Column bearing retainers have broken. Shaft or shaft enclosing tube has broken. Cavitation due to low submergence or operation of pump beyond maximum capacity rating. Presence of foreign material in the pump.
7	Delivery of oily water	Over-filling of oil. Foaming because of improper grade oil. Leaks at fittings.

turbine pump. However, the various troubles and the suggested remedies are tabulated in Table 4.3.

4.12 Submersible pump

When a vertical turbine pump is closely coupled with a borehole diameter size submerged electric motor, it is termed as *submersible pump*. In this case, electric motor and pump element operate entirely under submerged conditions. Since, the power unit (i.e. motor) is closely coupled with the pump; there is no need of any vertical shaft. As, the pump element is same in the cases of submersible as well as in vertical turbine pump, the pump characteristics will remain same. However, the efficiency is increased due to the elimination of long vertical shaft in case of submersible pump. Few advantages of submersible pump are written below.

- It can be used up to the depths, which are beyond the practical limits of vertical turbine pump.
- The system has no working part aboveground. Therefore, these wells can be installed in the locations where aboveground infrastructure is not appropriate (such as open playgrounds).

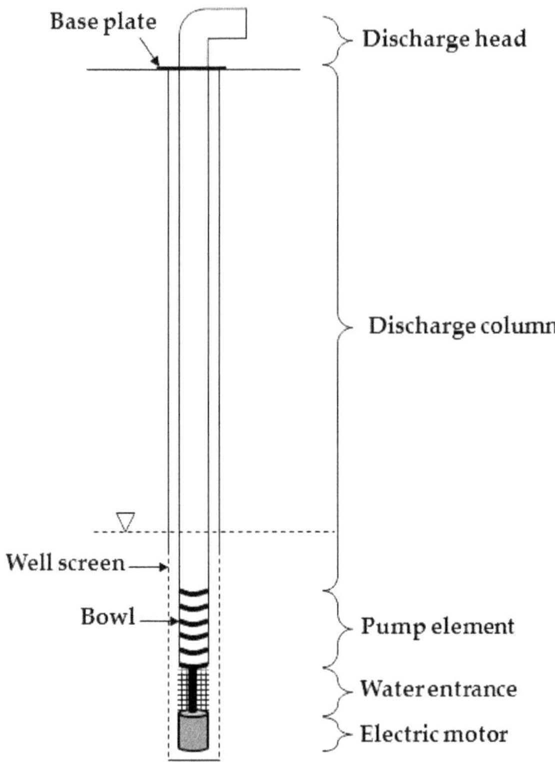

Figure 4.18: Schematic sketch showing the submersible pump and its elements.

Submersible pump can be divided into four parts, (i) *electric motor*, (ii) *pump element*, (iii) *discharge column*, and (iv) *discharge head* (Figure 4.18). The diameter of the *electric motor* is same as pump so that it can be conveniently accommodated in the well. It may be of dry or wet type. In dry type, the motor is enclosed in a steel case, which is filled with a light oil of high dielectric strength. Mercury is used as a seal, which prevents the movement of oil from inside of motor to outside and water from outside to inside motor. In wet type, water has access to every part accept the stator windings, which is completely sealed with a thin stainless steel inner liner. To prevent the entry of abrasive material into the motor, a filter is provided around the shaft.

A waterproof electric cable is used to transmit electricity to motor from the starter (at ground surface), which passes through the borehole (but outside the discharge column). The pump element is similar to vertical turbine pump. The water entrance is located between the motor and pump. Head assembly at the ground surface is provided for hanging the electric motor, pump element and discharge column. Generally, a pipe clamp is used for the purpose (Figure 4.19). For controlling the discharge rate, a sluice gate is provided at the end of discharge pipe. While lowering

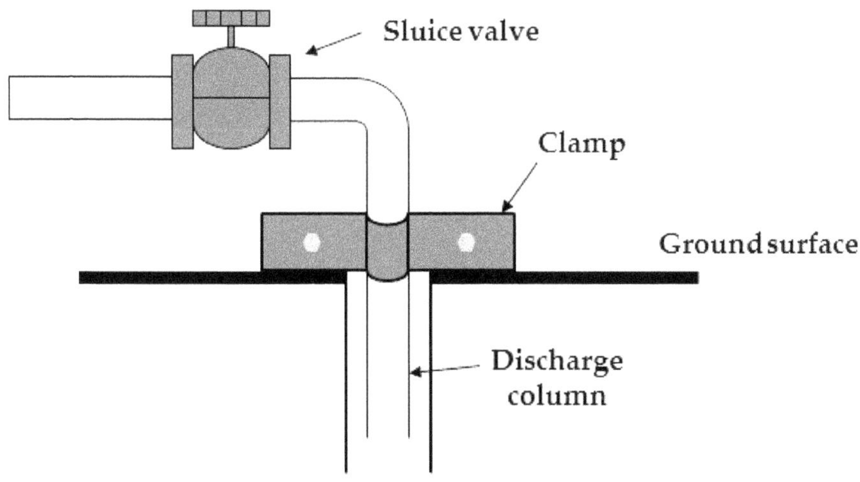

Figure 4.19: Schematic sketch showing the clamping system for submersible pump.

Table 4.4: Submersible pump troubles and checks.

S. No.	Troubles	General causes and checks
1	Tripping of circuit while motor is running	Input voltage may not be correct (check the line voltage with a voltmeter, which should be within maximum to minimum range). Starter or control has overheated from sunlight or any other source of heat (ventilate or shade the box). Control box components are defective (use an ohmmeter for check). Electric cable or motor winding may be defective (check the resistance of the motor winding by using an ohmmeter). Pump is overloading (if the line ampere reading is more than 5% above the manufacturer's provided value, the pump is overloading).
2	Motor is not starting even though fuses are not blowing	Input voltage may not be correct (check the line voltage). There may be no power (check power supply indicator). Control box may be defective (examine the wiring in the control box and tighten the contacts).
3	Pump is either delivering very little or no water	Possible reason is the airlocking of the pump (start and stop the pump several times giving at least 1-minute interval between each cycle). Line voltage may be low (check the line voltage with pump in running condition). Lowering of water level (lower the pump). Control valve on the discharge line may be partially/fully closed (examine it). Leakage in the discharge pipe (check and replace the section). Intake screen of the pump is blocked (sand or mud may be reasons). Parts of the pump have worn out (improper screening and pumping of sandy water may abrade the pump parts severely). Uncoupling of shaft between motor and pump (pull out the pump and coupled it properly).

the submersible pump, the electric cable is taped to the discharge column at every 2 m. During the initial run, pump is started with sluice gate closed or slightly open. Purpose is to check the presence of impurities in water (i.e. mud, sand, or other gritty particles). In case the pump is discharging water with impurities, care should

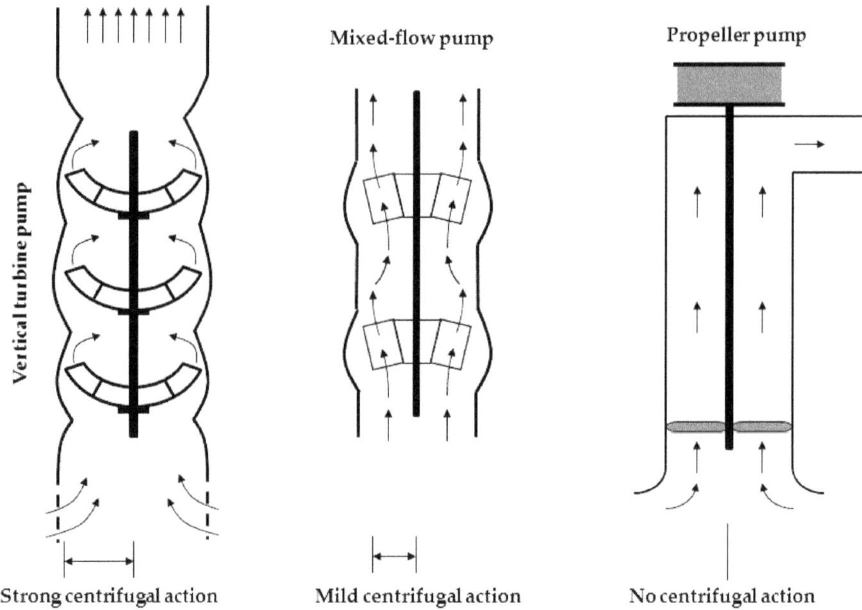

Figure 4.20: Schematic sketches showing the characteristic difference in the principle of operation of vertical turbine, mixed-flow, and propeller pumps.

be taken not to stop pumping. The pumping of the aquifer should be continued until the discharge of clear water. In the next step, the discharge rate of the pump is slightly increased by opening the sluice gate to little more extent. The pumped water is again checked for the presence of formation material. The mentioned step is repeated until the full opening of sluice valve. The basic purpose of starting the pump by keeping the sluice valve partially closed during the initial run is to slow down the sand movement. Excessive sand movement during the first few minutes of the initial run may wear out pump parts severely. It is necessary to overhaul the submersible pump after about two years of service life (i.e. around 6000 hours of operation). The various troubles and the suggested checks are tabulated in Table 4.4.

4.13 Propeller and mixed flow pumps

For pumping high discharge under low head conditions, propeller or mixed flow pumps are adapted. These pumps are specifically adapted for pumping water from canals, rivers, or streams. In Figure 4.20, comparison between the principles of operation of the turbine, propeller, and mixed flow pumps is shown. In turbine pump, pressure head is developed by the centrifugal force. In propeller pump, pressure head is developed by the propelling or lifting action of the vanes (i.e. by propeller). However, in case of mixed flow pump, head is developed partly by the centrifugal force and partly by the propelling action of vanes. Propeller pumps are known as *axial flow pumps* as flow through it is parallel to the axis of drive shaft.

Figure 4.21: Photograph of a propeller pump.

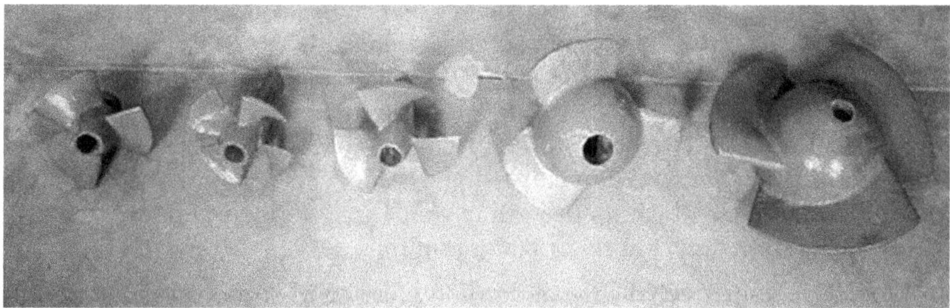

Figure 4.22: Photograph showing propellers of different sizes and different blade angles.

Propeller pumps are usually operated at much higher speeds than those of ordinary centrifugal pumps and are adapted to pump large volumes of water at comparatively low heads. The presence of suspended sediments or any other foreign material in water does not clog the propeller pump. These pumps can efficiently work within the pumping heads ranging from 1 to 2.5 m. The diameter of the cylindrical casing (or discharge column) of a propeller pump may range from 20 to 120 cm which depends upon the capacity of the pump required. To reduce the entrance losses, the lower portion of the pump is given a flared shape (Figure 4.21).

Water is lifted by the propeller having 3 to 5 blades. The angle of blades depends upon the working head and speed (Figure 4.22). Usually the blades (generally of

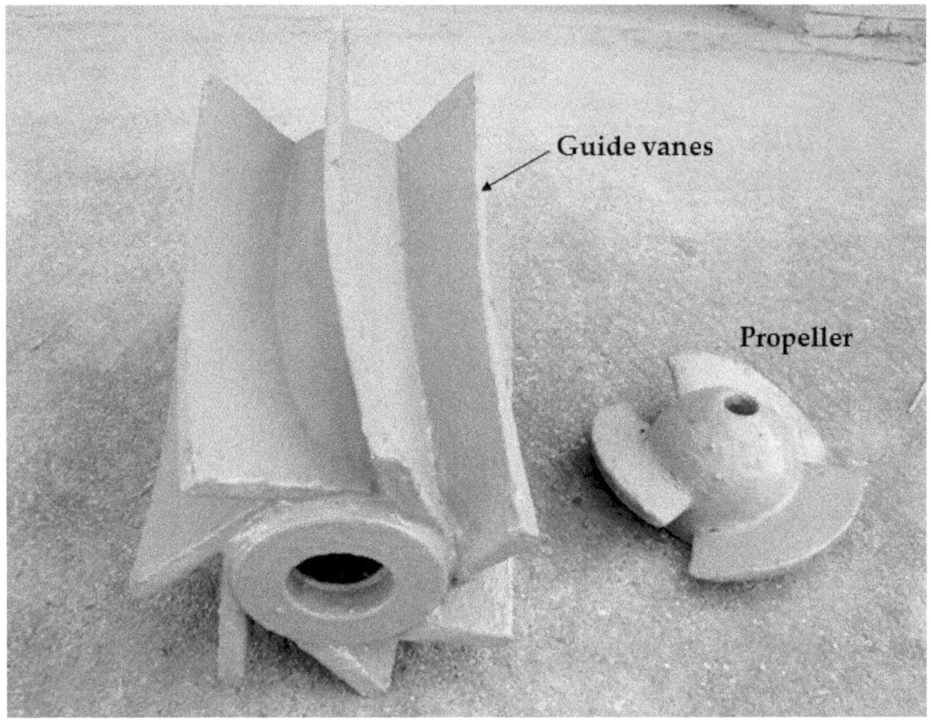

Figure 4.23: Photograph showing the guide vanes to be fixed above the propeller to smoothen out the water swirls.

bronze alloy) are casted to the central hub at fixed angles for fabricating propellers. However, the propeller with adjustable angle blades can be fabricated by adapting threaded joints. Above the impeller, guide vanes are provided to smoothen out the swirling action of rising column of water (Figure 4.23).

With the increase in head, the discharge capacity of the propeller pump is reduced drastically. The closing of discharge valve creates overloading conditions in the propeller pump and due to which steep horsepower curve is obtained. The selection of a propeller pump is not an appropriate decision under suction lift conditions. In case of propeller pump, the submergence depth is the distance of the lowest part of the suction pipe from the pumping water level. For the propeller pumps having size ranging from 20 to 35 cm, the minimum submergence required varies from 65 to 90 cm (Michael, 1997). The submergence depth of propeller pumps less than the minimum required results in the cavitation phenomenon. For the pump size range of 20 to 30 cm, the pump inlet should be at least 30 to 50 cm away from the sidewall of the water body (i.e. canal, river, or pond etc.). Moreover, the inlet should also be 20 to 30 cm above the bed of water body (Michael, 1997).

It is already mentioned that a *mixed-flow pump* represents the mixed features of both vertical turbine pump and the propeller pump. The working head is partly

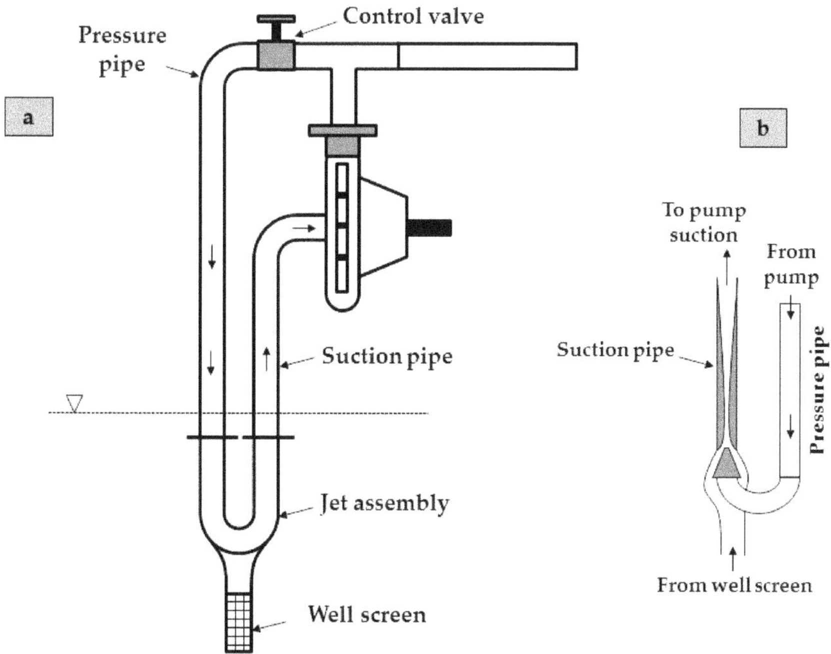

Figure 4.24: Schematic sketches showing (a) jet pump and (b) jet operation.

developed by the centrifugal force and partly by the propelling force (Figure 4.20). Within the head range of 5 to 10 m, this type of pump can lift water very efficiently. The impeller of a mixed-flow pump imparts outward thrust to water in addition to the velocity upwards. The flow is straightened by providing curved vanes above impeller. Usually, 2 to 3 stages are required for achieving the satisfactory performance.

4.14 Jet pump

A centrifugal pump when used in conjunction with jet mechanism (or ejector) is known as *jet pump* (Figure 4.24a). A considerable portion of high-pressure water delivered by the centrifugal pump is allowed to return through pressure pipe to activate the nozzle in the ejector. The reduction in pressure due to high velocity jet through venturi draws more water from the well (Figure 4.24b).

The control valve on the pressure pipe is used to maintain the necessary pressure to produce flow at the existing pumping head. Jet pump is not an efficient machine. It achieves maximum efficiency up to 35% only. The pump efficiency is mainly influenced by the nozzle-throat ratio. Jet pump can be conveniently installed in a well diameter as small as 5 cm. In comparison to ordinary centrifugal pump, high suction lift can be achieved which may usually range from 12 to 18 m for medium size jet pumps. The jet pump is not suitable in locations where, (i) water levels are

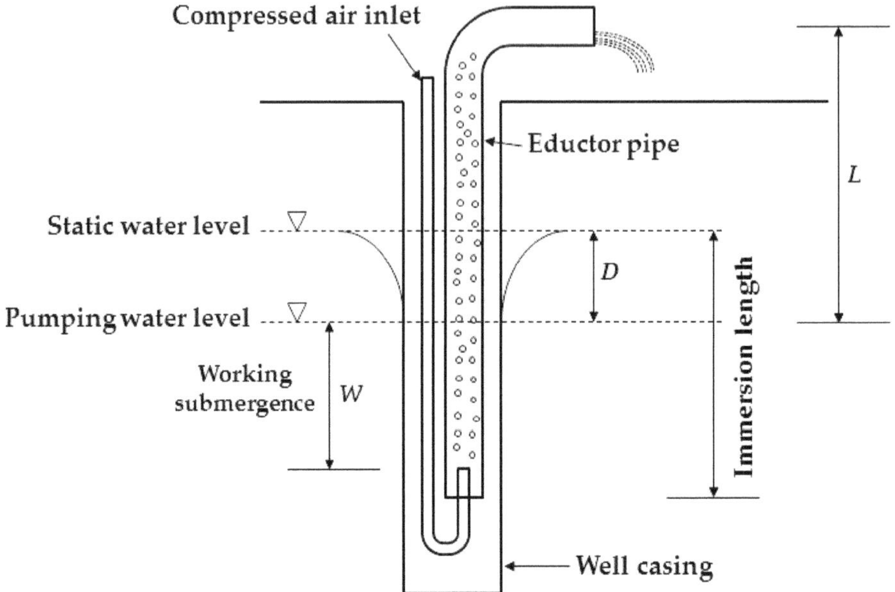

Figure 4.25: Schematic sketch showing the working principle of airlift pump.

subject to large variations, (ii) corrosion enlarges the nozzle, and (iii) incrustation plugs the nozzle.

4.15 Airlift pump

The working principle of the airlift pump is shown in Figure 4.25. When compressed air is blown into the eductor pipe's lowest point, a froth of air and water is produced which has a lower density than water and consequently rises to the surface. In simple terminology, we can say that air bubbles act as pneumatic pistons, which push or draw water up a pipe or stack as they rise and expand (Wurts et al., 1994). The major condition for efficient working of airlift pump is that both the eductor and air pipes must be submerged in water in the well with 50% or more of their lengths extending below the pumping level. Therefore, the borehole needs to be drilled to a depth more than twice the depth of the water level to allow adequate submergence. If the diameter of the well casing is not much larger than the air pipe, it can be used as eductor pipe. This particular method can be used to pump sand and drilling mud from the bottom of the well during its development operation.

The percent submergence and the relative sizes of the eductor and air pipes are two important factors in airlift pumping. From Figure 4.25, the percent submergence can be evaluated equal to

$$= \frac{W}{W+L} \times 100 \tag{4.13}$$

In the initial stage, the air discharge has to lift the water column of length equal to $(W+D)$. With the start of normal pumping, drawdown curve develops and the pumping water level dips down. Therefore, the working pressure also decreases. The formula for evaluating pressure from head

$$P = 0.0981 \times h \times \gamma \tag{4.14}$$

In Eq. (4.14), P is the pressure (in Kg/cm^2), h is the head (in m) and γ is the specific gravity of water. Using the formula (4.14), the starting and working pressures required can be written as

$$Starting\ pressure,\ P_s = \frac{W+D}{9.81} Kg/cm^2 \tag{4.15a}$$

$$Working\ pressure,\ P_w = \frac{W}{9.81} Kg/cm^2 \tag{4.15b}$$

The air injection should be regulated properly for achieving maximum efficiency. Too little air may reduce the yield whereas too much may cause excessive friction in the pipe. Other than groundwater pumping, airlift pumps are widely used by the aqua-culturists to pump, circulate, and aerate water in closed, re-circulating systems as well as in ponds.

4.16 Pump selection

Pump size can be decided based upon two factors *viz.* (i) the safe yield of the well, and (ii) the maximum water requirement of the crops in a day during peak demand periods. The dominating factor is considered for designing the same. While designing, it is essential to consider the efficiency of water conveyance system to the fields. Discharge capacity of a pump can also be evaluated based upon the crop requirements. It requires (i) area under different crops, (ii) water requirement of all the crops, (iii) interval between two successive irrigations of a crop (i.e. rotation period), and (iv) pumping hours in a day. The following formula is used to compute the discharge rate of required pump.

$$Q = 27.78 \times \frac{Ad_i}{RT} \tag{4.16}$$

In Eq. (4.16), Q is the pumping rate (l/s), A is the cropped area (ha), d_i is the depth of irrigation (cm), R is the rotation period (d), and T is the pumping duration in a day (h/d).

4.16.1 *Point of selection* – The selection of a pump for a particular well is influenced by its drawdown-discharge characteristics. The drawdown-discharge curve (or head-capacity curve) is placed over the pump characteristics curve. The point where the head-capacity curves of the pump and well intersects is considered as the *point of selection.*

4.17 Head loss due to friction in pipe systems

Head loss due to friction in a pipe flowing full – It can be evaluated using the formula as

$$h_f = \frac{4 f l v^2}{2gd} \tag{4.17}$$

In Eq. (4.17), f is the coefficient of friction of pipe (m), l is the pipe length (m), and d is the pipe diameter (m).

Head loss due to friction in a strainer – It can be evaluated using the formula as

$$h_f = K_s \frac{v^2}{2g} \tag{4.18}$$

In Eq. (4.18), K_s is a constant having value 0.95.

Head loss due to friction in a foot valve – It can be evaluated using the formula as

$$h_f = K_f \frac{v^2}{2g} \tag{4.19}$$

In Eq. (4.19), K_f is a constant having value 0.80.

5
Waterlogged Soils

5.1 Introduction

Waterlogging may be defined as the retention of cultivable land under water for a considerable period of time; which may cause severe damage or complete failure of crop. It is not necessary that water must stand on the ground surface to define as waterlogged soil. The areas, where water table rises to such an extent that the complete porous media in the crop root zone becomes saturated may also be considered as waterlogged soils.

Until date, there is no proper definition of waterlogging which could be commonly used throughout the world. Different organizations have defined the term waterlogging based on their mandates, which are written in the Table 5.1.

Depending upon the depth of water from the ground surface, different areas may be classified as

i Ponded/flooded/waterlogged lands – In this case, water table reaches above the ground surface and creates ponded/flooded conditions, which may persist for longer durations. According to Ministry of Water Resources, River Development & Ganga Rejuvenation, if water table is within 2 m of the land surface it may be considered as waterlogged area.

ii Potential area for waterlogging – When water table is between 2 to 3 m below land surface, it may be considered as potentially waterlogged.

Table 5.1: Various definitions of waterlogging.

Name of the organization	Definition
The National Commission on Agriculture	An area is said to be waterlogged when the water table rises to an extent that the soil pores in the root zone of a crop becomes saturated resulting in restriction of the normal circulation of air, decline in the level of oxygen and increase in the level of carbon dioxide.
Ministry of Water Resources, River Development & Ganga Rejuvenation	An irrigated area is said to be waterlogged when the surplus water stagnates due to poor drainage or when the shallow water table rises to an extent that soil pores in the root zone of a crop become saturated, resulting in restriction of the normal circulation of the air, decline in the level of oxygen and increase in the level of carbon dioxide.
Central Ground Water Board (CGWB)	It is that hydrologic condition when groundwater table reaches the shallow critical and semi-critical conditions and soil profile becomes partially or fully saturated with capillary groundwater.

In these areas, capillary rise may reach up to ground surface. Since, the capillary rise depends upon the soil type, it will vary in different soils.

iii Safe area – When water table is at least 3 m below the land surface, it may be considered as safe from waterlogging point of view. In these areas, the capillary rise may reach up to the level little below the crop root zone.

5.1.1 *Effects of waterlogging* – The filling of water in the pores of soil displaces the air and obstructs the release of gases given off by the crop roots. Under waterlogged conditions, the gas exchange is limited to top few centimeters of soil and free oxygen is practically absent below that. With the advent of waterlogged conditions, oxygen in the soil and in the water disappears within few hours. Subsequently, anaerobic decomposition of organic matter takes place and which is responsible for the reduction of mineral substances in the soil. After submergence of soil, toxic concentrations of ferrous and sulphide ions develop within few days; whereas, manganous ions take little longer time to do so. Under waterlogged conditions, rate of decomposition of organic matter slows down, which leads to the locking up of nitrogen in the organic residue. Therefore, the availability of nitrogen for normal growth of crop in poorly drained soils is a limiting factor. The yellowing, reddening, or scorched/strippled appearance of the leaves of the plants under waterlogged conditions indicates an unbalance in nutrient supply (Luthin, 1973).

Under anaerobic conditions, decomposition of organic matter leads to the release of methane and other harmful compounds, which is responsible for inhibiting the growth of few crops (e.g. tomato). High moisture content in the soil may be a cause of its compaction by machinery. Low temperature severely affects the seed germination. Plant diseases (specifically fungus) attack more actively under higher moisture or waterlogged conditions. Lack of proper drainage in irrigated areas may convert fertile lands in to salt affected lands.

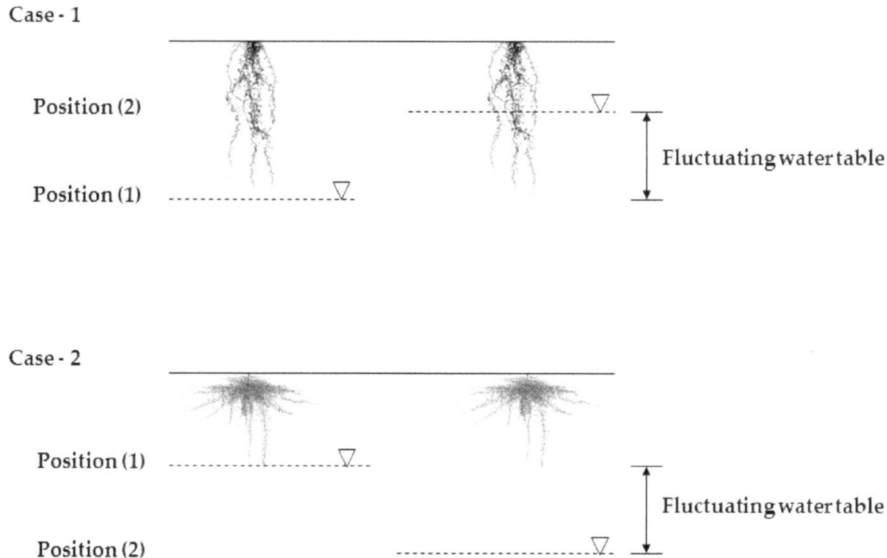

Figure 5.1: Schematic sketches showing the effect of fluctuating water table on crop growth.

5.1.2 *Effect of water table fluctuation* – Fluctuation of water table from near the ground surface to some depth or vice-versa cannot be considered beneficial for crop growth. It is explained through Figure 5.1. In case – 1, originally water table was at position (1) and due to the deeper water table the roots are developed more in the vertical direction than lateral. With time, the water table rises to position (2) and the roots of the crop are submerged. Consequently, crop roots rot off. In case – 2, originally water table was at position (1) and due to shallow water depth, roots are developed more in the lateral than vertical direction. With the lowering of water table to position (2), the moisture availability to the crop ceases. Therefore, the fluctuating water table in both the conditions severely affects the crop production.

5.1.3 *Drainage benefits* – By draining the soils, scalding (i.e. which may occur under waterlogged conditions in summer) of the crops can be avoided. Standing water on the soil is a breeding ground of mosquitoes and a health hazard. Adequate moisture content in the soil ease the farming operations. Adequately drained soil warm up readily and promotes germination. A good root environment exists in areas of adequately drained soils.

5.1.4 *Observing depth of water table* – Monitoring and plotting the depths of water in an area is the first-ever information that is required to identify the waterlogged areas. It can be accomplished by digging a hole in the soil and observing the level to which the water rises in the hole. These holes are commonly known as *observation wells.* An auger hole is the simplest form of an observation well (Figure 5.2). Depending upon the conditions of the soil, it may be cased or uncased.

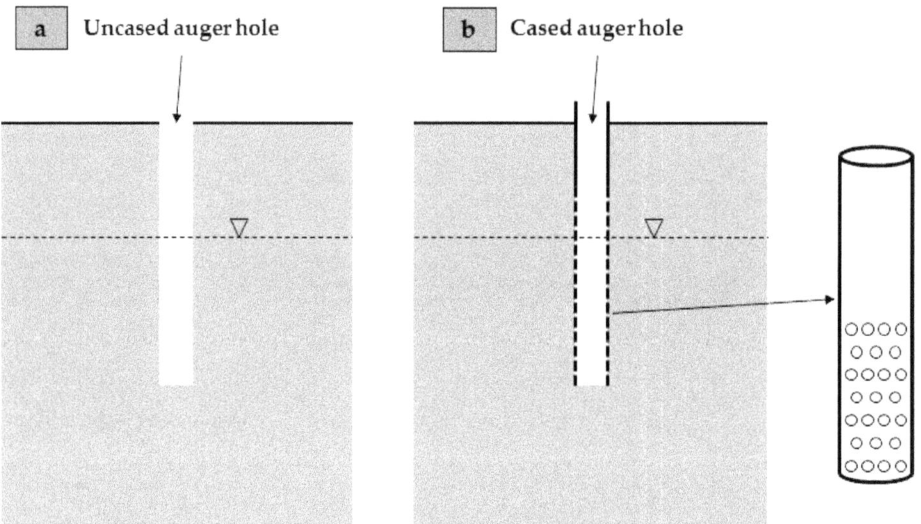

Figure 5.2: Schematic sketches showing the (a) uncased auger hole, and (b) cased auger hole.

5.1.5 *Observation of artesian conditions* – Sometimes, the waterlogged conditions are created by the artesian aquifers having high water pressures. It is difficult to drain such areas. Therefore, it is necessary to detect the exact cause of waterlogging before installing a drainage system. The information regarding the artesian pressures can be determined by (i) examining existing artesian wells in the area, (ii) observations noted by the well drillers, and (iii) any other locally available information (Luthin, 1973).

If no information is available for any area, a series of piezometers can be installed for generating authentic data. Unlike observation well (which is a perforated pipe), piezometer is an open-ended pipe which is installed in the aquifer to know the water pressure at the point of penetration. The pressure head at the point of penetration is the height to which water will rise in the pipe. Usually, piezometers are installed side by side, each at different depth. By knowing the pressure head values of different piezometers, vertical gradients between them are evaluated. If the water in the deeper piezometer rises to greater elevation than the water in the shallower piezometer, it indicates the artesian pressure (Figure 5.3a). In general, PVC pipes of diameter up to 50 mm are used as piezometers. With the use of a jetting rig, piezometer at any depth can be installed. For precise reading, piezometer pipe should be sealed off with the formation materials (specifically the confining layers of artesian aquifer) (Figure 5.3b). In case, the piezometer has to cross the gravel stratum, bentonite clay (or driller's mud) is used to seal off the gravel layer (Figure 5.3b).

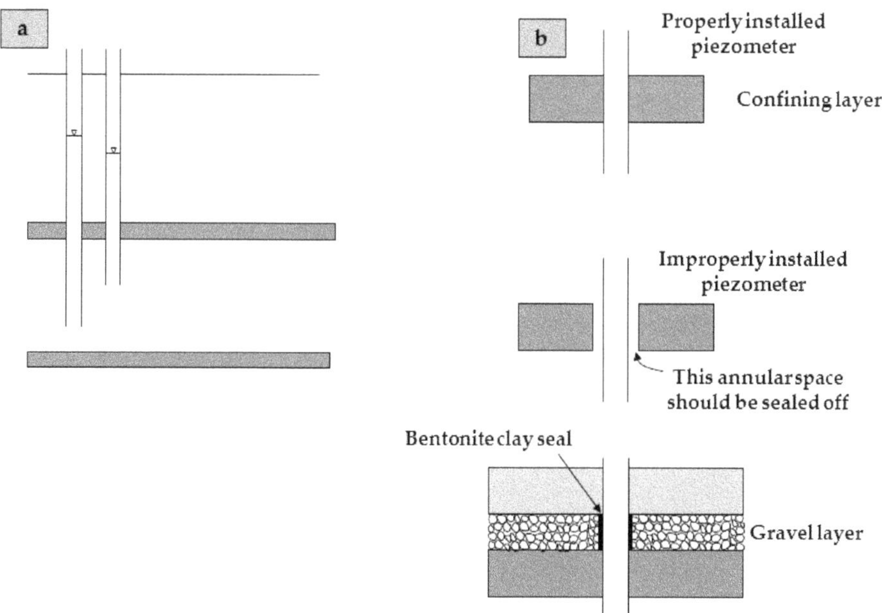

Figure 5.3: Schematic sketches showing the (a) installed piezometers, and (b) proper way of installing these.

5.2 Nature and extent of waterlogging problem

5.2.1 *The depth of impermeable layer* – For determining the depth and spacing of the drains, the information about the depth of the impermeable layer is required primarily. In the alluvial planes, impermeable layers are formed because of the stratification of the sediments that have been deposited from the streams.

In actual field conditions, impermeable layers (i.e. aquifuge) are very rare. To some or more extent, all the soil layers are permeable. From the drainage point of view, the definition of impermeability is decided on the relative variation in permeability. According to Luthin (1973), if the permeability of a particular layer is one-tenth of the adjoining layer, it can be considered as impermeable layer.

5.2.2 *Judging waterlogging problem from soil color* – It is a general practice to indicate the depth of water table during the survey of area. Experienced persons can judge the condition of the soil from its color. For example, if soil appears in orange and black mottling color, it indicates the periodic reducing (waterlogging) and oxidizing (lowering of water table) conditions of the ferric and manganic ions. Similarly, ferrous iron imparts blue color to the soil, which indicates the reduced conditions in the soil because of continuous submergence.

5.2.3 *Performance verification of an existing system* – By monitoring the outflow data of an existing drainage system, its *unit-area discharge* can be evaluated which can be further used to verify its performance.

5.2.4 *Topographic map* – The topographic map of the waterlogged area provides the information on the surface slopes, location, type of outlet, and the degree of land preparation if required. Therefore, this map should include the information on the position, alignment, and gradient of existing ditches; existing streams, culverts and other natural and artificial features that may influence the drainage system. The capacities and dimensions of natural channels should be determined.

6

Soil Permeability Measurement

6.1 Introduction

Permeability is a physical property of soil. For understanding the problems related to water seepage and drainage, permeability of the soil is required. In the present chapter, different methods (direct/indirect) of permeability measurement are described.

6.2 Methods

Soil permeability can be determined in two ways

- Directly, by passing water through the porous medium (either disturbed or undisturbed sample).
- Indirectly, by measuring those properties of the porous medium, which are related to permeability.

6.2.1 *Falling-head permeameter (direct method)* – In Figure 6.1, schematic sketch of a permeameter has been shown. As per the concept shown, water will move upward from the bottom of soil column to its top. Let Q_T is the total flow passed through the soil column. Then according to Darcy's law, flow rate per unit time can be evaluated as

$$\frac{dQ_T}{dt} = K \times \frac{H}{L} \times A \tag{6.1}$$

In Eq. (6.1), K is the hydraulic conductivity; H is the hydraulic head in the water column measured with respect to the upper surface of soil; A is the cross-sectional

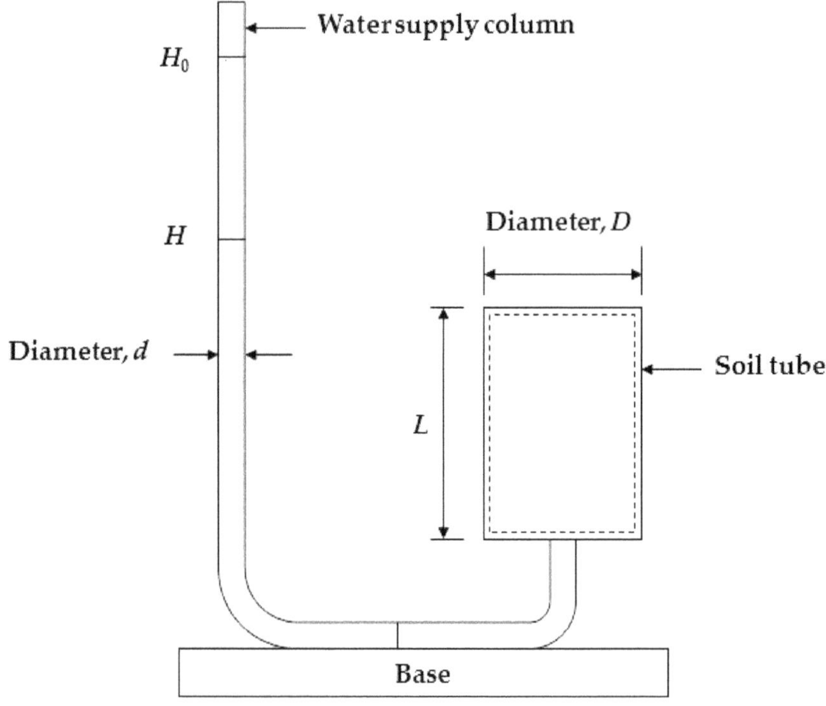

Figure 6.1: Schematic sketch showing the falling-head permeameter.

area of the soil column; and L is the length of the soil column. With the opening of valve, water from the supply column drops from the H_0 level to H in a time (t). Let 'a' is the cross-sectional area of the water supply column, then the total volume of water passed through soil column can be calculated as

$$Q_T = a\,H_0 - a\,H \tag{6.2}$$

Flow per unit head can be evaluated by differentiating Eq. (6.2) with respect to H as

$$\frac{dQ_T}{dH} = -a \tag{6.3}$$

Equating Eq. (6.1) with Eq. (6.3) for dQ_T and separating the hydraulic head terms, it rearranges to

$$-\frac{dH}{H} = \frac{KA}{aL}dt \tag{6.4}$$

On integrating the Eq. (6.4) from the limits as $H = H_0$ at $t = t_0$ and $H = H$at $t = t$, it results in

$$K = \frac{a\,L\log_e\left(\dfrac{H_0}{H}\right)}{A\left(t - t_0\right)} \tag{6.5}$$

By writing, the mathematical formulae for the terms '*a*' and *A*, Eq. (6.5) further simplifies to

$$K = \frac{d^2\,L\,\log_e\left(\dfrac{H_0}{H}\right)}{D^2\left(t - t_0\right)} \tag{6.6}$$

In Eq. (6.6), *d* is the diameter of the water supply pipe and *D* is the diameter of the soil column. The measured value is multiplied by the temperature correction factor to bring it to standard value at 15.5°C.

6.2.2 *The Kozeny-Carman equation* (*Indirect method*) – The basic equation was developed firstly by Kozeny (1927), which was later reviewed and extend by Carman (1937) in the form as

$$v = \frac{\rho g}{k_c\, n\, S_v^2} \times \frac{n^2}{1 - n^2} \times i \tag{6.7}$$

In Eq. (6.7), *v* is the velocity flux; k_c is the Kozeny-Carman constant, whose value for unconsolidated materials is empirically evaluated equal to 5.0; *n* is the medium porosity; S_v is the surface area per unit volume; and *i* is the hydraulic gradient.

6.2.3 *Childs-Marshall equation* – The equation is based on the Poiseuille's law (i.e. hypothesis of capillary-tube). A method for calculating the soil permeability from the pore size distribution was given by Marshall (1957) as

$$k = \frac{\varepsilon^2}{n^2}\left[\frac{\left(r_1^2 + 3r_2^2 + 5r_3^2 + 7r_4^2 + \cdots + (3n-1)r_n^2\right)}{8}\right] \tag{6.8}$$

In Eq. (6.8), *ε* is the porosity of the porous medium (cm³/cm²); *k* is the permeability in cm²; and $r_1, r_2, r_3, \ldots.. r_n$ (cm) represent the mean radius of pores (in decreasing order of size) in each of *n* equal fractions of the total pore space.

6.2.4 *Soil textural vs. K relationship* – A correlation between soil texture and hydraulic conductivity (from laboratory study) can be established (Aronovici, 1947). This method is suitable where soil permeability is controlled by soil texture rather than soil structure; but this method is not successful for salt affected soils.

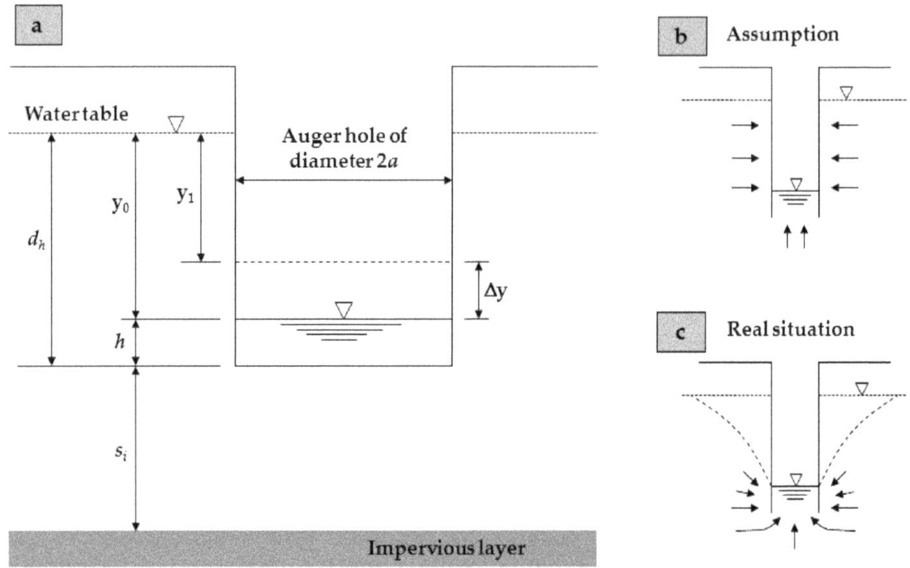

Figure 6.2: Schematic sketches showing (a) hydraulic conductivity measurement by single-auger hole method, (b) assumption for deriving mathematical equation, and (c) the real situation.

6.3 In situ measurement of hydraulic conductivity in presence of water table

In this method, single auger hole is dug into the ground below the water table. The benefits of this method are (i) the presence of rock or root holes adjacent to the auger hole does not affect the accuracy of method, (ii) the measurement largely reflects the horizontal component of the conductivity, which is an important component from drainage point of view, and (iii) in situ conditions.

6.3.1 *Single-auger hole method* – The variables involved in the single-auger hole method are shown in Figure 6.2a. After digging the hole, the level of water in it is allowed to come in equilibrium with the original water table. After attaining the equilibrium conditions, the water is pumped out of the hole and the rise in water level is monitored with time.

For evaluating hydraulic conductivity from the monitored transient water level data, a mathematical relation is required. Hooghoudt has derived the equations for two conditions *viz.* (i) when the depth of auger hole reaches up to an impervious layer, and (ii) when impervious layer is at great depth from the end of auger hole. Assumptions made by Hooghoudt while deriving these equations are written below

- Water table is not lowered around the auger hole when water is pumped out of it (Figure 6.2b). In actual practice, a drawdown curve is established due to the difference in original water table and pumped water level in hole.
- It is considered that water flows into the auger hole horizontally from the sides and vertically from the bottom (as is shown in Figure 6.2b). However,

in real situation, the water entry into the auger hole is not according to the said assumption (Figure 6.2c).

Based on above assumptions, the filling rate of auger hole is proportional to its circumference and is inversely proportional to its cross-sectional area. The relation in mathematical terms can be written as

$$\frac{dy}{dt} = -K\frac{2\pi a d_h}{\pi a^2} \times \frac{y}{S} \tag{6.9}$$

In Eq. (6.9), $S = \frac{ad_h}{0.19}$ and has dimensions of length. The relation for upward movement of water through the base of auger hole can be written as

$$\frac{dy}{dt} = -K\frac{y}{S} \tag{6.10}$$

Therefore, for evaluating total flow into the auger hole, the Eq.s (6.9) and (6.10) are added and the resulting equation can be written as

$$\frac{dy}{dt} = -K\frac{(2d_h + a)y}{S} \tag{6.11}$$

Integrating Eq. (6.11) between the limits from $y = y_0$ at $t = 0$ to $y = y_1$ at $t = t_1$, it changes to

$$K = \frac{2.303\,aS}{(2d_h + a)\Delta t}\log_{10}\frac{y_0}{y_1} \tag{6.12}$$

For the second case when auger hole terminates at the impervious layer, the flow from the base of the hole ceases and the Eq. (6.12) modifies to

$$K = \frac{2.303\,aS}{2d_h\,\Delta t}\log_{10}\frac{y_0}{y_1} \tag{6.13}$$

Incorporating the value of S in Eq. (6.13), it changes to

$$K = \frac{2.303\,a^2}{0.38\,\Delta t}\log_{10}\frac{y_0}{y_1} \tag{6.14}$$

6.3.2 *Two-auger hole method* – The method was first proposed by Childs and Collis-George (1950) for the non-layered soils. In this method, two holes (instead of one) of equal diameter and depth are dug out. The experimental set-up has been shown in Figure 6.3.

This method is valid only when holes are dug up to impermeable layer. Water is pumped from one hole and conveyed to other. By doing so, a small hydraulic head difference is created between the levels of water in the two holes. From the geometry of Figure 6.3, the hydraulic conductivity (K) is given by

$$K = \frac{Q}{\pi L \Delta H} \cosh^{-1} \frac{b}{2a} \tag{6.15}$$

Further, they have also suggested the end and capillary fringe corrections as 20 and 5 cm, respectively. The suggested values are added to L for correct results.

6.3.3 *The pipe-cavity method* – In this method, a pipe (or a tube) is inserted in the soil below the water table. Then soil is augered out of the tube and pipe is raised a little to form a cavity at the end of tube (Figure 6.4).

Otherwise, a slightly larger diameter pipe can be inserted in the auger hole leaving the lower L length portion for a cavity. After the establishment of equilibrium conditions water is pumped out of the tube. Kirkham (1946) has proposed a relation between the water rising rate in the tube with the hydraulic conductivity of the soil, which can be written as

$$K = \frac{\pi r^2 \log_e \left(\frac{y_0}{y_1} \right)}{S(t_2 - t_1)} \tag{6.16}$$

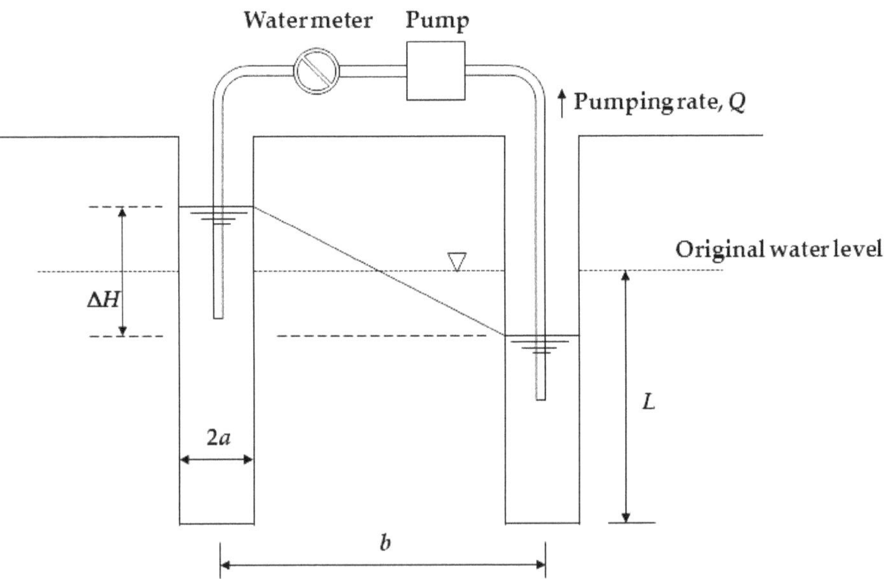

Figure 6.3: Schematic sketch showing hydraulic conductivity measurement by two-auger hole method.

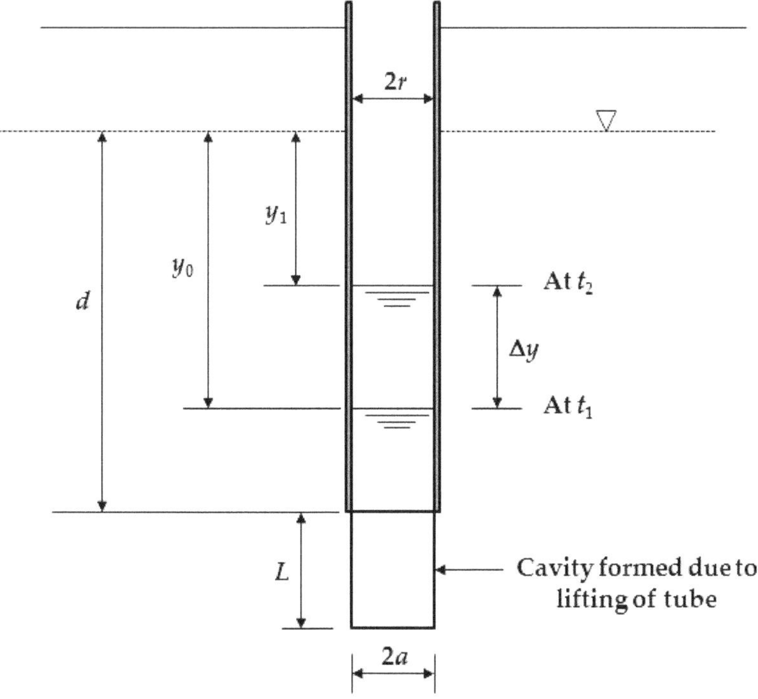

Figure 6.4: Schematic sketch showing hydraulic conductivity measurement by pipe-cavity method.

In Eq. (6.16), y_0 and y_1 are the distances from water table to water levels in tube or pipe at times t_1 and t_2, respectively; r is the radius of pipe; S is a coefficient determined by Kirkham with the electric analogue. Interestingly, the existence of cavity is immaterial for using Eq. (6.16).

6.4 In situ measurement of hydraulic conductivity in the absence of water table

6.4.1 *Shallow well pump-in test* – The method is schematically explained in Figure 6.5. After digging a hole to the desired depth, a constant head of water is maintained in it. The use of water from the tank is monitored with time. The experiment is carried-on until the constant rate is obtained. The requirement of long period (generally 2 to 6 days), use of large quantity of water, and involvement of considerable amount of equipment are principal limitations of this method.

6.4.2 *Pond-infiltration test* – To avoid the soil compaction that is inherent in the core sampling methods, pond-infiltration test is suggested. A soil dike is formed around a circular area (with 4 m diameter). This area is kept filled with water until the saturated conditions are achieved. Then, falling water level is observed with time. Since, the flow of water inside the soil is entirely due to gravity, a unit hydraulic

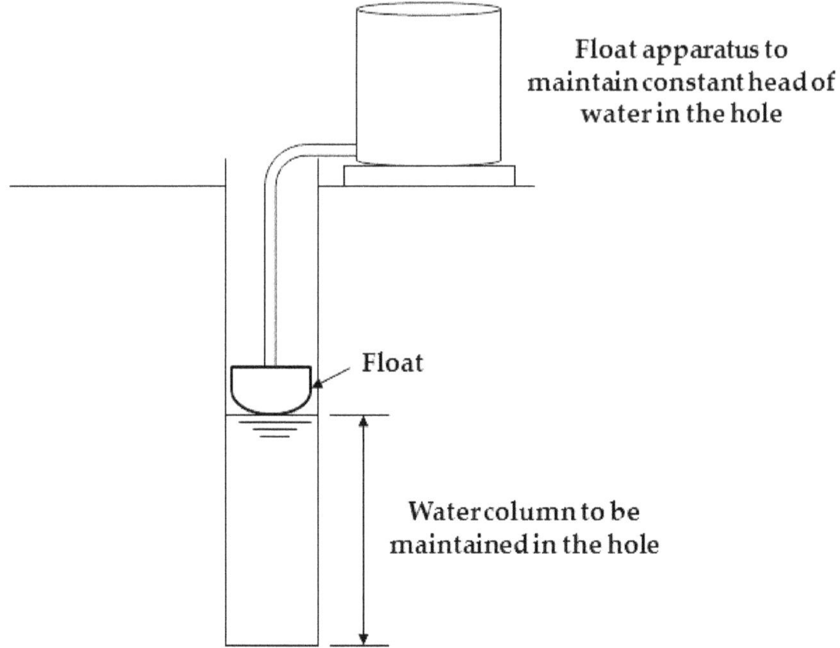

Figure 6.5: Schematic sketch showing hydraulic conductivity measurement by shallow well pump-in test method.

gradient can be considered. From the available data, hydraulic conductivity can be easily evaluated using Darcy's law as

$$K = \frac{Q}{A} \quad as \;\; hydraulic \;\; gradient, \; i = 1 \tag{6.17}$$

7

Equations for Evaluating Depth and Spacing of Drains

7.1 Introduction

Subsurface drainage system controls the depth of water table and salinity level in the root zone by draining the excess water in the desired zone. The efficiency of any drainage system depends upon its proper installation. Its performance is influenced by the drain properties (which includes its diameter, depth, and spacing), soil characteristics (its profile and hydraulic conductivity), water table depth, and the desired discharge. Based upon certain assumptions, the mentioned parameters can be grouped into various mathematical formulations, which are presented in this chapter.

7.2 Hooghoudt's Equation

The water table position in an area having subsurface drainage system depends upon the following factors.

- Hydraulic conductivity of soil
- Depth and spacing of drains
- Rainfall/irrigation rate
- Depth of impermeable layer with reference to the bottom of drain
- Plant intake, deep percolation, and soil stratification (usually ignored)

Figure 7.1 shows a schematic sketch as analyzed by Hooghoudt.

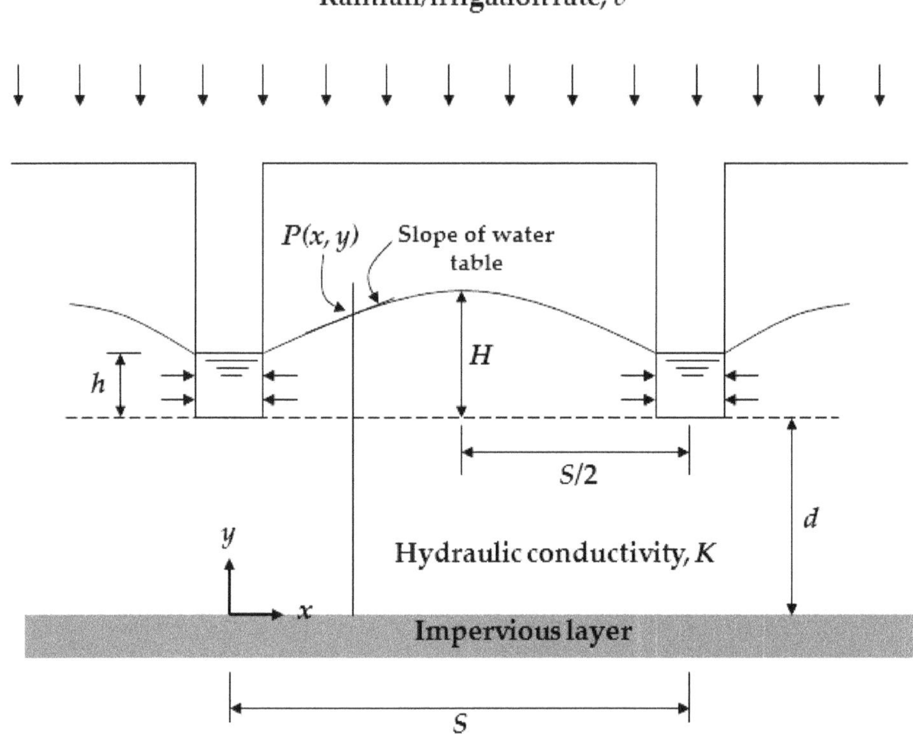

Figure 7.1: Schematic sketch showing the elements of Hooghoudt's drain spacing formula.

The analysis is based on the assumption that rainfall or irrigation rate (i.e. v) is constant. For simplifying the mathematical development, it is assumed that the hydraulic gradient at any point in the flow region is equal to the slope of water table above that point (see Figure 7.1). The said assumption is known as Dupuit-Forchheimer assumption, which is the basic assumption for studying the groundwater flow. This assumption implies that groundwater flows horizontally as equipotential lines are vertical (Figure 7.2a). However, the perusal of Figure 7.2b shows that this assumption is not valid near the drains.

As per the variables shown in the Figure 7.1, Hooghoudt equation can be written as

$$S^2 = \frac{4K\left(H^2 - h^2 + 2d\,H - 2d\,h\right)}{v} \tag{7.1}$$

According to Eq. (7.1), the drain spacing S becomes infinite as d goes to infinity. It is because of the Dupuit-Forchheimer assumption, which does not account for the radial flow into the drain. Equation (7.1) for empty drains changes to

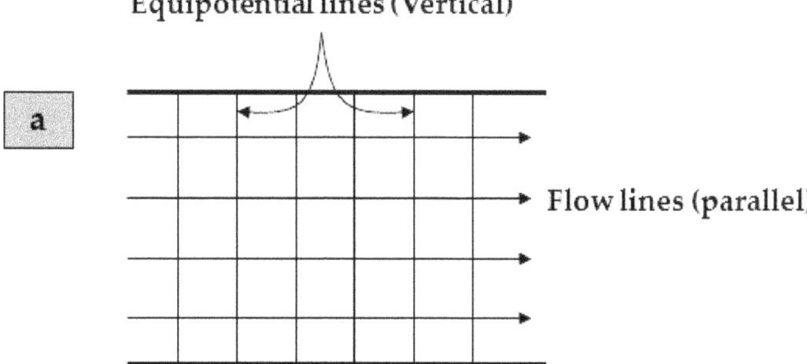

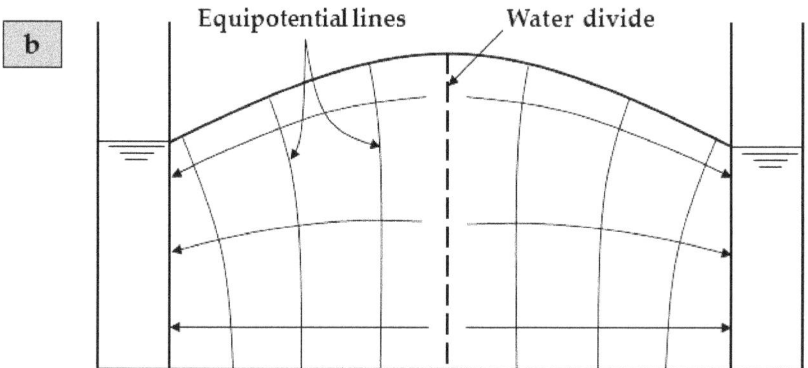

Figure 7.2: Schematic sketches showing (a) horizontal flow through porous media and (b) the flow towards the drains.

$$S^2 = \frac{4KH(H+2d)}{v} \tag{7.2}$$

Equation (7.2) is an ellipse equation with semi-major and semi-minor axes as $\frac{S}{2}$ and $\frac{S}{2}\sqrt{\frac{v}{K}}$, respectively. Therefore, the Hooghoudt's equation is also known as *ellipse equation* and sometimes *Donnan equation.*

7.2.1 *Hooghoudt's equation for a layered soil* – The Hooghoudt's equation can be written for layered soil with reference to the drain line, above which hydraulic conductivity is K_a and below that K_b (Figure 7.3).

The Hooghoudt's equation for the layered soil as per the variables shown in the Figure 7.3 can be written as

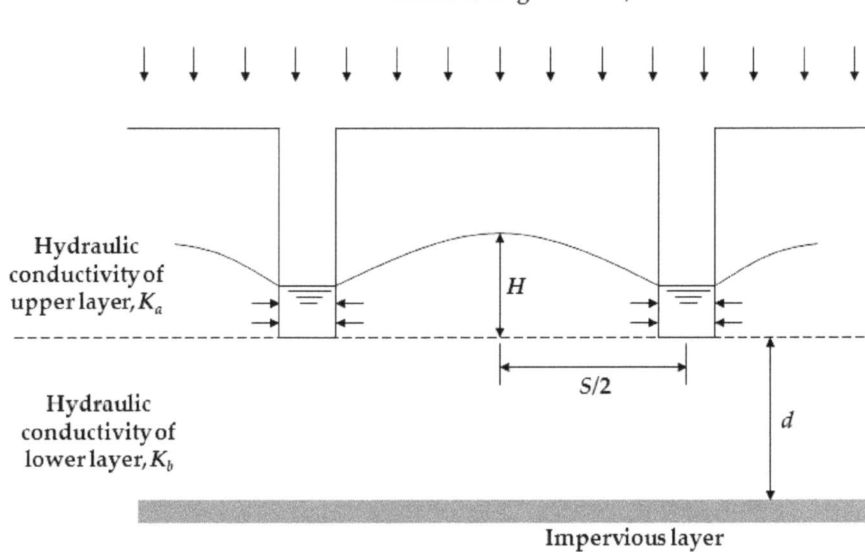

Figure 7.3: Schematic sketch showing the elements of Hooghoudt's drain spacing formula for layered soil.

$$S^2 = \frac{4}{v}\left(K_a H^2\right) + \frac{8}{v}\left(K_b d H\right) \tag{7.3}$$

The presence of layers above the drain line are given proper weight-age in the following way

$$K_a = \frac{K_1 l_1 + K_2 l_2 + \ldots + K_i l_i}{l_1 + l_2 + \ldots + l_i} \tag{7.4}$$

In Eq. (7.4), K_i is the hydraulic conductivity of the ith layer having thickness l_i. If a layer below the drain level has permeability one-tenth of the upper layer, it can be safely considered as impermeable layer (or barrier layer for Hooghoudt equation). For making the Eq. (7.1) valid for larger *d*, Hooghoudt developed a table of equivalent depths.

7.2.2 *Factors affecting the replenishment rate of water table* – The factors that influence the replenishment rate of water table are

- Initial moisture content of the soil at start of the rainfall/irrigation.
- Interception losses, which further depends on the type of crop.
- Downward movement of water due to deep seepage losses through the barrier layer.

- Upward movement of water from deeper aquifer having high artesian pressure.
- Escape of water from the area due to surface runoff, which depends on the infiltration rate of the soil.

7.3 Kirkham formula

With reference to the conditions shown in Figure 7.4, a relation to find depth and spacing of drains is given by Kirkham (1958) which was based on exact mathematical procedures as

$$H^d = \frac{2Sv}{K} \cdot \frac{1}{\pi} \left[\log_e \frac{2S}{\pi r} + \sum_{m=1}^{\infty} \left\{ \frac{1}{m} \left(\mathrm{Cos} \frac{m\pi r}{S} - \mathrm{Cos}\, m\pi \right) \times \left(\mathrm{Coth} \frac{m\pi h}{S} - 1 \right) \right\} \right] \quad (7.5)$$

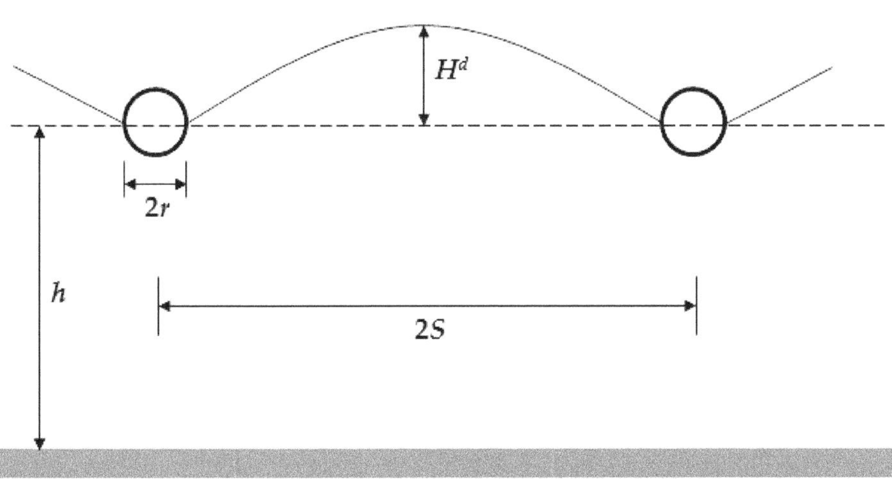

Figure 7.4: Schematic sketch showing the elements of Kirkham equation.

In Eq. (7.5), H^d is the maximum height of water table above the drain level; v is the rainfall/irrigation rate; K is the hydraulic conductivity; h is the height of water table immediately over the drain level from the impermeable layer; $2S$ is the drain spacing; and r is the radius of pipe.

7.4 Bureau of reclamation formula

Based upon the principle that moving water table (transient) will provide a more accurate method of design, Bureau of Reclamation has given formula as

$$H = \frac{192}{\pi^3} \sum_{n=1,2,3}^{\infty} (-1)^{\frac{(n-1)}{2}} \left\{ \frac{\left[\left(n^2 - \frac{8}{\pi^2} \right) \right]}{n^5} \right\} \cdot e^{\left(-\frac{\pi^2 n^2 \alpha t}{L^2} \right)} \quad with\ \alpha = \frac{KD}{s} \tag{7.6}$$

In Eq. (7.6), H is the water table height midway between the drains; K is the hydraulic conductivity; D is the average depth of groundwater stream; L is the drain spacing; s is the specific yield; and t is the time. The above solution is based on the assumption that water table initially (i.e. $t = 0$) has a shape of fourth degree parabola whose equation can be written as

$$y = \frac{8H}{L^4}\left(L^3 x - 3L^2 x^2 + 4Lx^3 - 2x^4\right) \tag{7.7}$$

As per Bureau of Reclamation, when annual recharge in an area is equal to annual discharge, this condition may be defined as *'dynamic equilibrium'*.

7.5 Drainage of areas having artesian conditions

For draining areas having considerable artesian conditions, the following relation is used to design the drainage system.

$$H = -\varphi_a \left[\frac{\log_e \left\{ \frac{\left(\text{Cosh} \frac{2\pi r}{S} - 1 \right)}{\left[\left(\text{Cosh} \frac{2\pi (r + 2h)}{S} - 1 \right) \right]} \right\}}{\frac{\log_e 2}{\left(\text{Cosh} \frac{4\pi h}{S} + 1 \right)}} \right] \tag{7.8}$$

In Eq. (7.8), H is the water table height at the midpoint between the drains; ϕ_a is the hydraulic head in artesian aquifer; r is the radius of drain; S is the spacing between the drains; and h describes the distance from tile line to artesian layer.

8
Subsurface Drains

8.1 Introduction

A drain beneath the surface of soil is known as *subsurface drain*. Before installing the subsurface drainage system, a choice between open surface, covered, tube well, and mole drainage is required. The mole and tube well drainage systems are generally installed under very specific conditions. Most likely, the choice has to make between open or covered drains. Open drains can receive overland flow directly. However, it is associated with disadvantages like (i) loss of land, (ii) interference with the irrigation system, (iii) splitting up of land into fragments, which hampers farming operations, and (iv) maintenance. On the other hand, covered drains do not interfere with the farming operations and there is no loss in cultivable area. The different types of materials used for subsurface covered drainage system are (i) clay pipes, (ii) concrete pipes, (iii) plastic pipes, (iv) gravel blanket, (v) fibrous wood materials, (vi) stones, (vii) bituminous fibrous materials, (viii) perforated/ slotted plastic pipes, and (ix) any other material that can be placed in the soil and has considerable service life.

8.2 Types of outlets for subsurface drains

Two types of outlets are used for farm drainage systems *viz.* (i) gravity outlets, and (ii) pumped outlets. The selection of a particular outlet depends upon the topography of area and permeability of soil.

8.2.1 *Down wells* – Wherever the formations such as fractured lava or porous limestone exist beneath the soil, down wells can be used. Drained water from the area after proper filtration is allowed to fall into well.

8.2.2 *Discharge into an open drain* – It is the most commonly used method for discharging drained water. The end-section of the subsurface drain should be blind and not less than 3 m in length (Figure 8.1). The outlet of the subsurface drain should be at least 0.3 to 0.6 m above the expected high water level in the natural stream or collector drain.

8.2.3 *Evaporation sumps* – As it is clear from the name, evaporation sumps are used where there is no natural drainage ways or streams to carry away the drainage water.

8.2.4 *Pump outlets* – Sometimes, the outlet level of subsurface/collector drains are below the main drain level (Figure 8.2). Therefore, it becomes necessary to discharge

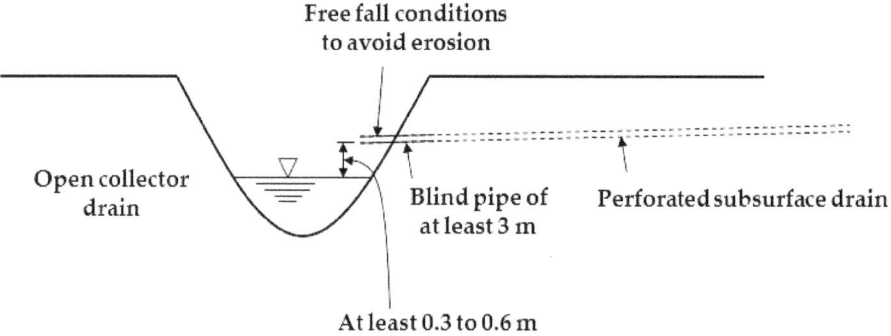

Figure 8.1: Schematic sketch showing the typical conditions for the end-section of subsurface drain.

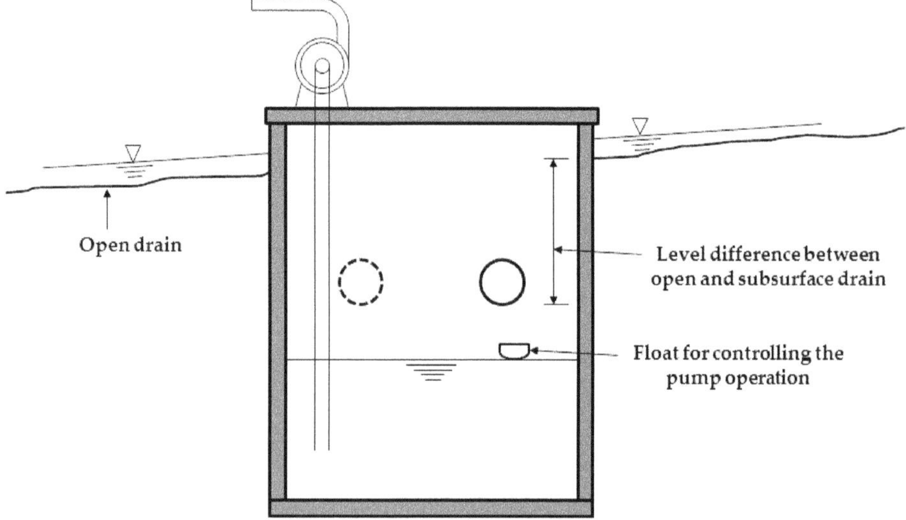

Figure 8.2: Schematic sketch showing the pump outlet for subsurface drainage system.

the water from the subsurface drains below main drain level into main drain. The most commonly used method is to construct a vertical sump (Figure 8.2). A float-controlled pump is used to empty the sump.

8.3 Hydraulic design of subsurface drains

8.3.1 *Drainage coefficient* – The amount of water removed from an area in 24 h period is known as drainage coefficient. The factors like rainfall rate and size of the drainage area affects this value.

8.3.2 *Slope of subsurface drain* – The self-cleaning of full flowing drain is possible if a flow velocity of 20 to 25 cm/s is maintained (Luthin, 1973). According to Frevert et al. (1955), a minimum grade of 0.15% for 4" pipe and 0.05% for 12" pipe is required for trouble-free service of drainage system.

8.4 Drain line protection

8.4.1 *Reasons of drain line failure* – The possible reasons are as follows

- Excessive crack width (in case of tiles) or excessively large size perforations/slots (in case of plastic pipe).
- Successive sections are not aligned properly.
- Collapse of drainpipe due to excessive load or inadequate strength.
- Misalignment of tile sections due to settlement of foundation or grade reversal in case of plastic pipe.
- Entry of backfill material into the drain line during compaction of backfill.
- Entry of fine sand and silt in the drain line due to improper design of gravel pack.

8.4.2 *Drain envelope* – The drain envelopes are used to restrict the entry of fine sand and silt into drain and variety of materials are used as drain envelopes. It includes properly designed gravel pack, coarse sand, and organic materials. The envelope should be designed in such a way that fine clay particles could pass through it. The criteria of gravel pack design are same as for an irrigation well.

8.5 Water entry into subsurface drain

In case of tile drains, water is entered through cracks (Figure 8.3a); whereas, in case of slotted pipes (Figure 8.3b), the entire permeable perimeter is available for water entry. By doubling the crack width, the water entry through cracks increases by 10%. For installing tile drains in unstable soils, the crack width should be as small as possible. The necessary space for water entry into the drain is automatically created by the irregularities of the segment ends of the pipe. In other words, each joint should be *less-than-perfect* fit. It is because of this purpose the joining segments of the tiles are not given proper finish while fabricating. The recommended crack width in the stable soils ranges from 3 to 6 mm. Maximum head loss occurs in the vicinity of drain. According to Luthin and Haig (1972), doubling the drain diameter from 2 to 4 inches increased the flow through it by 35 to 60%. They further reported

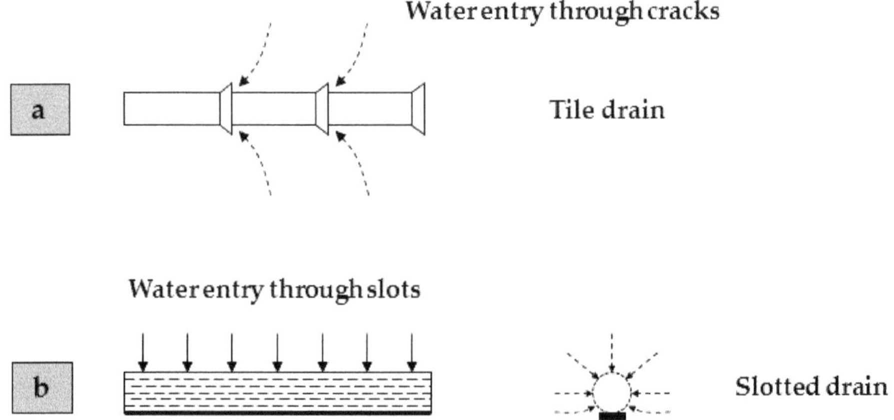

Figure 8.3: Schematic sketches showing the water entry in (a) tile drain, and (b) slotted drain.

that the flow rate through the pipe increased by 2.5 times if the pipe-segment length decreased from 3-feet to 1-foot.

8.5.1 *Effect of drain spacing and depth of impervious layer* – According to Kirkham's solution, the flow rate into a drain is independent of the spacing of the drains as long as they are more than 6 m apart. However, the flow into the drains is dependent upon the depth of impervious layer if its depth is not more than 60 cm. Moreover, with the decrease in depth of the impervious layer from the drain level, inflow rate decreases.

8.6 Effect of total area and location of openings on the inflow rate to drains

While experimenting on 100-mm diameter corrugated plastic pipe, Mohammad and Skaggs (1983) have found that the effective radius of the pipe increases with increase in perforated area. They reported that the effective radius increased from 5 mm to about 21 mm with increase in perforated area from 38 to 75 cm^2/m. However, they have found that the location of the drain perforations had little impact on the effective radius of the drainpipe. They further reported that the effective radius of the standard tube with 38.5 cm^2/m opening area was increased from 5 mm to 36 mm with the use of a 50 mm thick gravel envelope. It is well-established fact that with increase in the size of perforations, percent increase in inflow decreases (Kirkham and Schwab, 1951).

8.7 Formulations for evaluating loads on drain pipe

The method of laying the pipe in the bottom of a trench is referred to as the *bedding* of the pipe. The choice of the proper bedding procedure may increase the resistance of pipe against crushing. Depending upon the width of the trench, the loads on the drainpipe can be determined by the two formulas *viz.* (i) ditch conduit load formula, and (ii) projecting-conduit formula.

8.7.1 *Ditch-conduit load formula (for narrow trench)* – This formula is used for narrow trenches. The bulk density of the fill material is less than the adjoining soil (Figure 8.4a). During the settling of the filled material, there is a resistance between it and the walls of the trench. Consequently, the load on the drain due to weight of soil decreases. Marston (1930) has given this formula as

$$W_c = C_d \, w B_d^2 \tag{8.1}$$

In Eq. (8.1), W_c is the total load on the pipe; C_d is the load coefficient; w is the unit weight of the fill material; and B_d is the ditch width at the top of the conduit.

8.7.2 *Projecting-conduit formula (for wide trench)* – If the width of the ditch is more than 2 to 3 times the diameter of pipe, projecting-conduit formula is used (Figure 8.4b). In this case, the friction between filled material and the walls of the trench is not important. Moreover, the settling of the fill material is more on either side of the pipe than above the pipe. As a result, load on the pipe is increased. The relation in mathematically terms can be written as

$$W_c = C_c \, w B_c^2 \tag{8.2}$$

In Eq. (8.2), B_c is the outside diameter of the pipe; and C_c is the load coefficient.

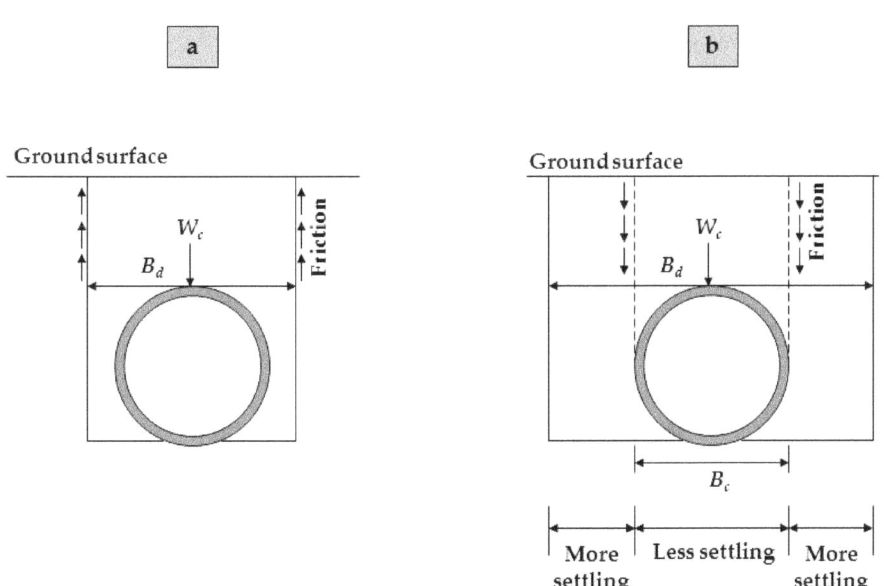

Figure 8.4: Schematic sketches showing the bedding of drainpipe in (a) narrow trench, and in (b) wide trench.

8.8 Blocking of drains due to chemical deposits

Bacterial activities oxidize and precipitate reduced forms of iron and manganese in the drainage water. Excess deposit may block the pipe and make it unserviceable. The oxides of iron and manganese can be dissolved by treating the deposits with 1N H_2SO_4 and 2% Sodium bisulphate solution (Spencer et al., 1963).

8.9 Installation of subsurface drains

Classically, the installation of subsurface drains includes marking the alignments and levels, excavation of trenches, placement of subsurface drainpipe as well as envelope material, and backfilling. However, these days the installation is completely based on machines and the steps followed while installing subsurface drains are

- Marking the alignments and levels
- Excavating the trenches and laying of drain pipes
- Collector installation
- Special considerations
- Supervision and inspection of drain installation

8.9.1 *Marking the alignments and levels* – Most adopted methods of marking alignments and levels is by placing stakes in the soil at both ends of a drain line, with the top of the stakes at a fixed height above the future trench bed. The slope of the drain line is thereby indicated implicitly. A row of boning rods is placed in line (both vertically and horizontally) between the stakes. The boning rods are thus in a line parallel to the trench bed, and grade control can be achieved through sighting by the driver of the drain line laying machine (Figure 8.5). The same principle can be applied when drains are installed manually.

8.9.2 *Excavating the trenches* – The trenches can be excavated manually or using machines.

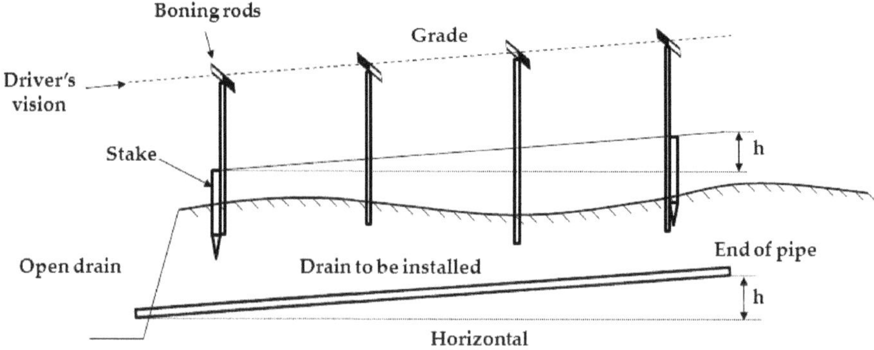

Figure 8.5: Schematic sketch showing the staking out for grade control of drainpipe.

8.9.2.1 *Manual digging* – The hand digging of drains is restricted to a depth of 1.5 m. Cost of the labor is the limiting factor for digging the deeper trenches.

8.9.2.2 *Digging of trenches using machinery* – The machines for installing subsurface drains can be classified into two main categories *viz.* (i) trenchers, and (ii) trenchless machines.

i Trenchers – The digging implement is commonly a continuous chain with knives. The excavation depth and trench width of a machine can be varied through inter-changing digging attachments. Clay tiles and concrete pipes move down a chute behind the digging chain.

ii Trenchless drainage machines – Trenchless drainage machines are in use since about 1965, after flexible corrugated plastic pipes appeared on the market. Two types of trenchless devices are *viz.* (a) vertical plough, and (b) V-plough.

a) Vertical plough – It acts as a sub-soiler. By this plough, the soil is lifted and large fissures and cracks are formed (Figure 8.6). If these extend down to the drain depth, the increased permeability leads to a low entrance resistance and an enhancement in flow of water into the pipe. However, beyond certain depth, the soil is pushed aside by the plough blade, instead of being lifted and fissured. This results in smearing, compaction, and a destruction of macro-pores, so that the permeability is reduced and the entrance resistance is increased. The critical depth depends mainly on the soil texture and on the moisture content during pipe installation.

b) V-plough – It lifts a triangular beam of soil and if the plough is not properly adjusted has a hazard of deforming the corrugated pipe

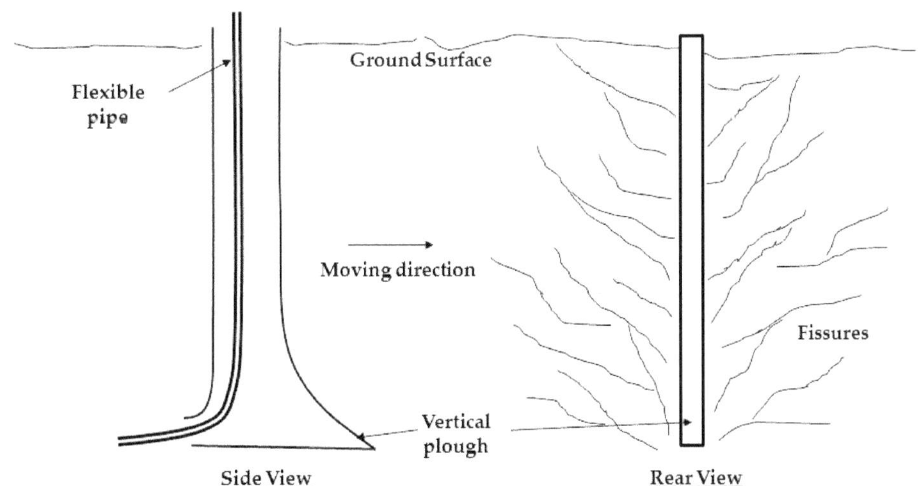

Figure 8.6: Schematic sketch showing the trenchless pipe drain installation using vertical plough.

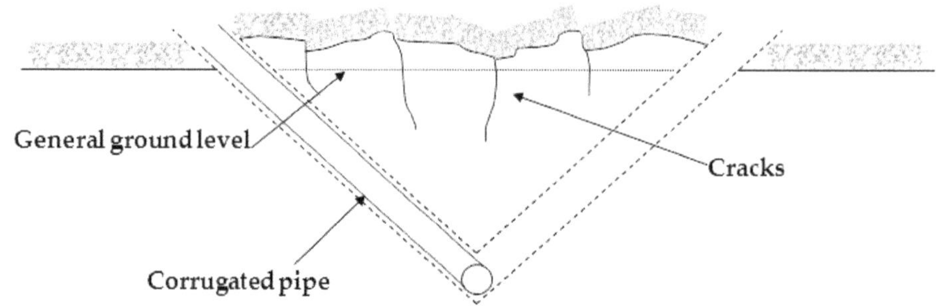

Figure 8.7: Schematic sketch showing the trenchless pipe drain installation using V-plough.

under the weight of the soil beam (Figure 8.7). We cannot use hard pipes with trenchless machine. Corrugated plastic pipes are the only feasible pipes for trenchless machines. A V-plough can handle a maximum outside pipe diameter (including the envelope) not more than 0.10 – 0.125 m, whereas vertical plough can handle much larger diameter.

8.9.3 *Collector installation* – Concrete pipes are installed either by trencher or by hydraulic excavator. As a safeguard against siltation, the collector is commonly laid as a closed conduit. Thus, the joints between pipe sections are sealed with either mortar or close fitting rubber rings. Installation of a deep collector in an unstable soil well below the water table is a difficult job, as sloughing conditions are faced during pipe installation. Installation of concrete pipes is possible only after the lowering of water table. This condition can be achieved by the well-pointing technique or alternatively by horizontal dewatering.

8.9.4 *Special considerations while installing subsurface drains* – A variety of adverse field conditions may jeopardize the laying of pipe drainage system if no special precautions are taken. Most of these hazardous conditions can be grouped as wet conditions and the most prevalent are

- High water table conditions.
- High water level in the open main drain.
- Waterlogged top soil due to recent irrigation or rainfall.
- Pipe-drain crossing an irrigation canal.

Sometimes, an appropriate choice of season may overcome many of these problems. However, following are the recommendations under above conditions.

- A general principle of drain installation is to start laying it from the downstream end, so that any free water can be drained away immediately. In addition, the water level in the open drain should be below the pipe outlets, and the connection with the collector should be made before a field drain is installed.

- The dry season should be selected for the installation of pipe drainage system.
- When open canals have to be crossed, these should be dry during pipe installation. It is advisable to lay a closed pipe section at the crossing to avoid piping from the canal seepage.
- At the outlet of a pipe drain, there is an extra risk of erosion of the trench backfill. As a precaution, the last few meters of the drain before the drain outlet should consist of un-perforated pipe without envelope material (Figure 8.1).
- Whenever trees or shrubs are growing near drainage pipe lines, there is a risk of root penetration into the pipes. Roots of perennial vegetation may grow into a drainpipe. They enter the pipe through the inlet openings, and cases are known in which they subsequently fill the entire drainpipe over a considerable length and thus seriously obstruct water flow. Recommended precautions are to keep drain lines away from trees wherever possible; and to use un-perforated pipe where strips of trees of shrubs must be crossed.

8.9.5 *Supervision and inspection of drain installation* – It is required for the following reasons.

- Most importantly, for handling the unforeseen conditions during installation.
- To ensure that the installation is according to the design specifications.
- Quality of the material being used.
- To ensure good workmanship.
- To ensure that the trenches are properly backfilled and compacted.
- To assess and justify any extra work or modification that is required for the efficient working of the system.

8.10 Subsurface drainage systems

Various subsurface drainage systems in practice are

- Natural or random
- Herringbone
- Gridiron
- Interceptor

8.10.1 *Natural or Random drainage system* – The method is most suitable for small or isolated wet areas (Figure 8.8). In general, the main drain follows the largest depression in the field and the submains and laterals are laid to drain the isolated wet areas.

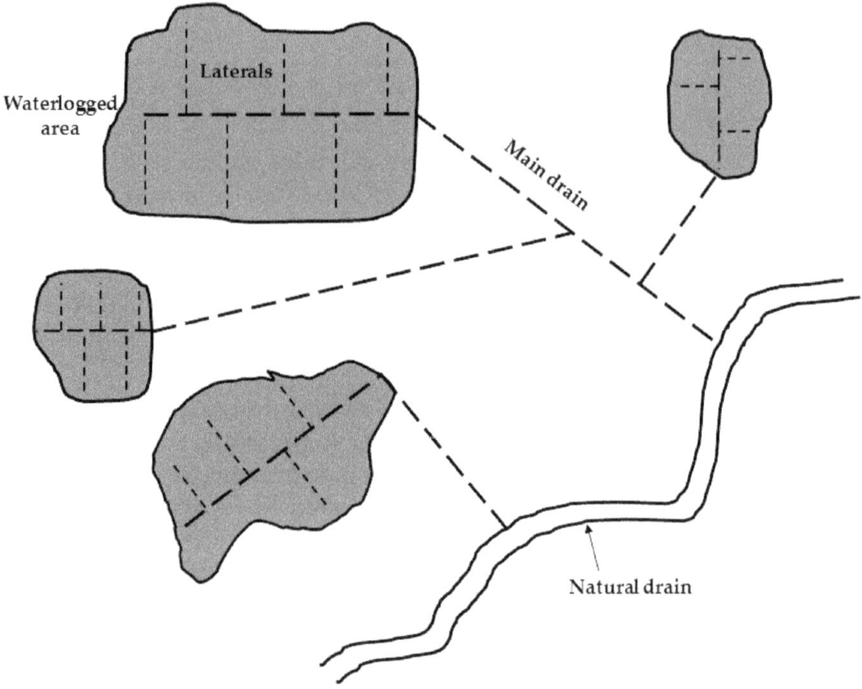

Figure 8.8: Schematic sketch showing the natural or random subsurface drainage system.

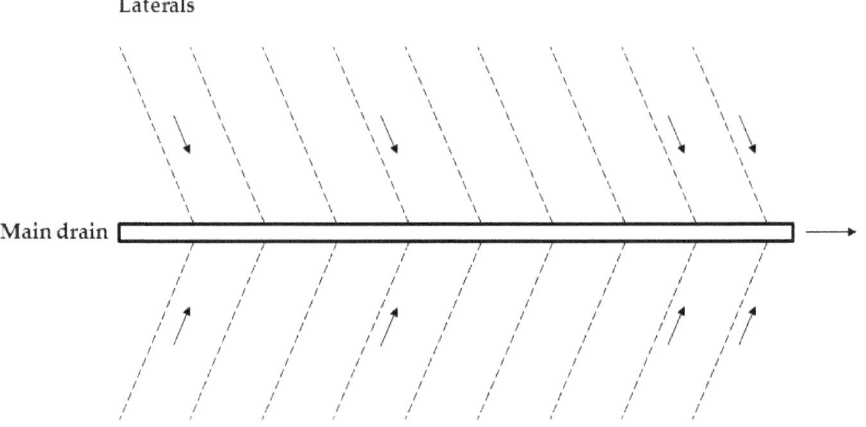

Figure 8.9: Schematic sketch showing the herringbone subsurface drainage system.

8.10.2 *Herringbone drainage system* – In this system, the laterals are laid along the slopes of the ground at convenient spacing and joined on either side of the main drain that follows the depressions in the direction perpendicular to the slope (Figure 8.9). The laterals are joined to the main at an angle.

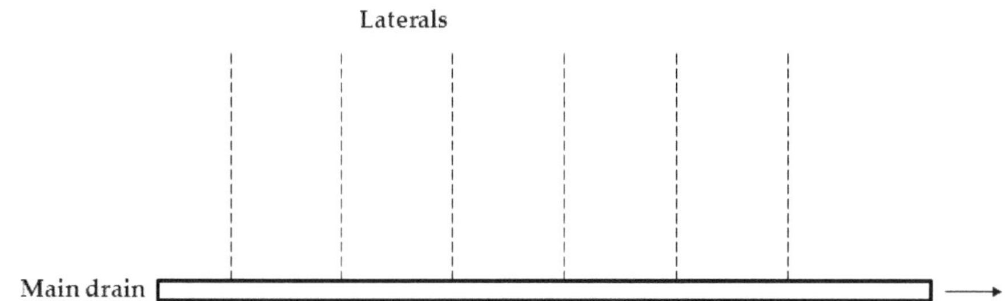

Figure 8.10: Schematic sketch showing the gridiron subsurface drainage system.

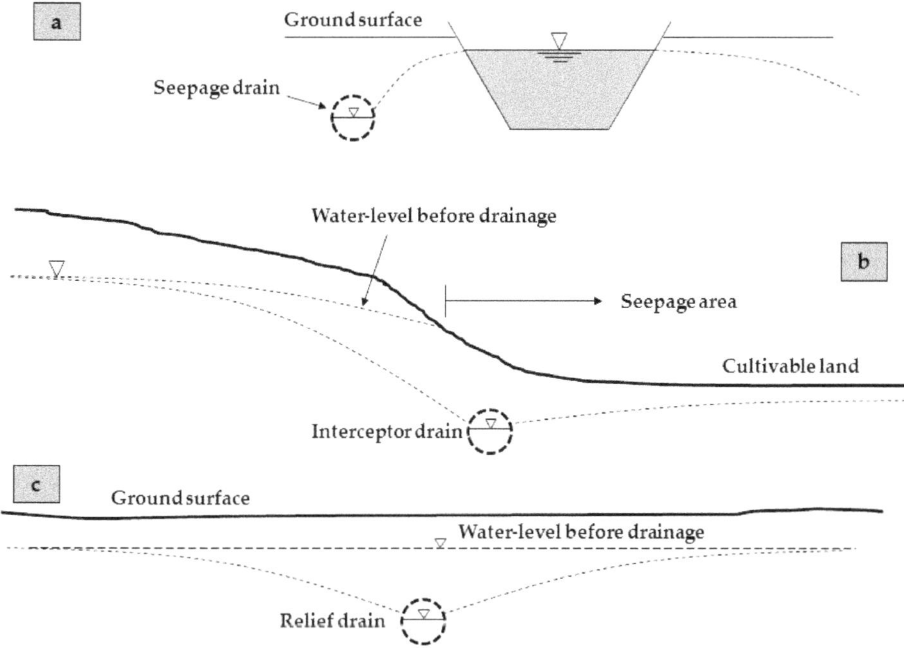

Figure 8.11: Schematic sketches showing the (a) seepage drain, (b) interceptor drain, and (c) relief drain.

8.10.3 *Gridiron drainage system* – In the gridiron system, laterals are laid along the slope of the ground and joined on only one side of the main drain (Figure 8.10). The system is ideal for the flat regularly shaped fields with uniform soil.

8.10.4 *Interceptor drain* – It is also known as *cut-off* or *seepage* drain. It is aligned along and adjacent to the source of seepage (Figure 8.11a). The interceptor drain intercepts the surface/subsurface flow from the higher reaches (Figure 8.11b). For differentiating purpose, it is worth to mention here that the relief drains are laid parallel to the direction of groundwater flow (Figure 8.11c).

8.11 Maintenance of subsurface drains

A mechanical rod made of flexible vanadium steel is used to remove plugging due to plant roots and soil materials in the drains. Blockage due to soil can also be removed by high-pressure stream of water by entering the hose from outlet end.

9
Tube Well Drainage Systems

9.1 Introduction

Tube well drainage consists of pumping from a single or series of wells, subject to the condition that they have capacity to pump groundwater equal to drainable surplus. In Indus plain of India, multiple well points system was used for draining waterlogged alkaline/sodic soils. Later-on this system become popular for skimming freshwater layers, which are floating on the brackish layers in an aquifer (Vashisht and Shakya, 2016a-b). For draining unconfined aquifers, gravity wells are used. Undoubtedly, wells are the only solution for reclaiming waterlogged conditions created due to artesian pressure in the semi-confined aquifer. It is worth to mention here that the existence of confined aquifers is very rare.

9.2 Comparison of tube well drainage with other subsurface drainage systems

In subsurface drainage system, open ditches or pipes are used to drain excess water from the soil profile of 1 to 3 m. Drained water is then conveyed to outlet through collector drains or allowed to directly fall into natural drain. On the other hand, tube well drainage system removes excess water from the aquifer depending upon its depth. The pumped water is then discharged into the natural drainage channel.

9.2.1 *Advantages of tube well drainage system* – Most suitable on undulating topography. Water table can be lowered to much greater depth than the other surface/subsurface drainage systems. In case the pervious substrata are found at a depth of 5 m or more, maximum benefits can be derived from tube well drainage system (Ritzema, 1994). If the groundwater quality is good, it can be used for irrigation by conveying it to the desired locations. In comparison to other drainage systems, the required length of open surface drains is considerably less.

9.2.2 *Disadvantages of tube well drainage system* – It is difficult to maintain a well than a pipe drain. For operating the well, energy in the form of electricity or diesel is required continuously. Constant withdrawal of water may lead to pressure reduction (or decline in water table) in extensive portion of the aquifer. This reduction in pressure (or water table decline) may lead domestic wells to dry. System is not suitable for lowering water table in small areas as water from the adjoining portions invade to the drained portion and tries to maintain the level. For lowering the water table instantly, dense network of wells is required (which raises the cost of operation). Tube well drainage is successful only if transmissivity of the aquifer is high (i.e. more than 600 m^2/d). The system is not successful under excessive seepage and artesian conditions.

9.3 Physical and economic feasibility of tube well drainage system

The physical and economic feasibility of the tube well drainage system depends upon the hydro-geological conditions of the area, which are discussed in the following sections.

9.3.1 *Thickness and hydraulic conductivity of the aquifer* – It is already mentioned in the section 9.2.2 that the tube well drainage system is not economically viable in aquifers with transmissivity lesser than 600 m^2/d. If we know the hydraulic conductivity of the aquifer, then the required minimum thickness of the formation can be evaluated.

9.3.2 *Hydraulic resistance of leaky layer* – Hydraulic resistance of a leaky layer plays an important role while using tube well drainage system. According to Ritzema (1994), the value of hydraulic resistance of the leaky layer much larger than 1000 days may not be considered feasible.

9.3.3 *Quality of groundwater* – Groundwater quality is an important criterion while deciding the fate of a tube well drainage system. If the groundwater quality is marginal to good, it can be conveyed to locations where irrigation is required. If groundwater quality is poor to brackish, it has to be disposed-off safely without leaving any opportunity for it to remixes with the groundwater. There are examples, when the major drainage/skimming projects have failed, as the disposal of non-usable water was not properly planned (Ali et al., 2004).

9.4 Pattern of well field

In general, more than one-well are used in tube well drainage system and these are installed in a particular pattern. If the number of wells in a well field are less than six, then depending upon the number of wells, triangular, square, or rectangular patterns are used. However, circular pattern is preferred when the number of wells is more than six. The number of wells in the circular pattern may vary from 6 to 12. Using the principle of superposition (as discussed in section 2.11), it is very easy to study the drawdown behavior of well fields. Therefore, the well hydraulics of single well is required. In this book, the derivation of the most general cases under steady state conditions (i.e. well in unconfined/confined aquifer) are discussed. Few solutions are already presented in the "well hydraulics" chapter. For latest solutions, students may consult relevant journals.

9.4.1 *Design of drainage well* – Dupuit has assumed that flow towards the fully penetrating well is horizontal. During pumping under steady state conditions, the flow towards the well in unconfined aquifer can be written using Darcy's equation as

$$Q = 2\pi rhK\frac{dh}{dr} \tag{9.1}$$

All the variables of the Eq. (9.1) are depicted in Figure 9.1.

After integrating the Eq. (9.1) using the boundary conditions *viz.* at $r = r_w$, $h = h_w$ and at $r = R_e$, $h = H$, it changes to

$$\frac{Q}{\pi K}\log_e\frac{R_e}{r_w} = H^2 - h_w^2 \tag{9.2}$$

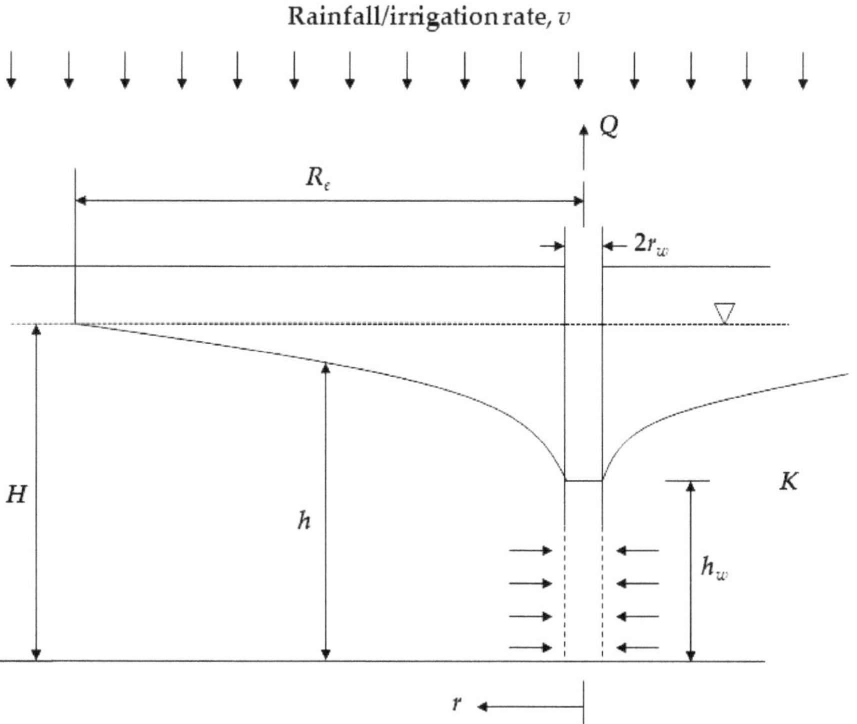

Figure 9.1: Schematic sketch showing the well installed in an unconfined aquifer.

Generalizing the Eq. (9.2) with the limit that there exists a head h at any radial distance r from the well, it modifies to

$$\frac{Q}{\pi K}\log_e\frac{r}{r_w} = h^2 - h_w^2 \tag{9.3}$$

Combining Eq. (9.2) with Eq. (9.3) with reference to Q, equation for free water surface (also called *Dupuit solution*) can be written as

$$h^2 = \frac{H^2 - h_w^2}{\log_e\frac{R_e}{r_w}}\left(\log_e\frac{r}{r_w}\right) + h_w^2 \tag{9.4}$$

Now, consider that aquifer is being replenished by the vertical flow (i.e. rainfall, irrigation from top and artesian seepage from below) along with the horizontal flow. Say, r is the radius of area from where vertical flow is coming to the aquifer. Total volume of water entering the aquifer (considering v is the steady state rainfall/ irrigation rate) will be equal to $\pi\left(r^2 - r_w^2\right)v$. Therefore, the equilibrium equation (after neglecting the well radius) can be written as

$$Q_r = Q - \pi v r^2 \tag{9.5}$$

In Eq. (9.5), Q_r is the resultant flow in the aquifer. Moreover, the resultant flow equation in the aquifer can also be written using Darcy's law as

$$Q_r = 2\pi r h K\frac{dh}{dr} \tag{9.6}$$

Combining the Eq. (9.5) with Eq. (9.6) with reference to term Q_r, we got

$$Q - \pi v r^2 = 2\pi r h K\frac{dh}{dr} \tag{9.7}$$

Integrating the Eq. (9.7) within the limits as $r = r_w$, $h = h_w$ and $r = R_e$, $h = H$, it changes to

$$\frac{Q}{\pi K}\log_e\frac{R_e}{r_w} - \frac{v\left(R_e^2 - r_w^2\right)}{2K} = H^2 - h_w^2 \tag{9.8}$$

Generalized form of Eq. (9.8) can be written as

$$\frac{Q}{\pi K}\log_e\frac{r}{r_w} - \frac{v\left(r^2 - r_w^2\right)}{2K} = h^2 - h_w^2 \tag{9.9}$$

On neglecting radius of well r_w with respect to radius of influence R_e, the term $\left(R_e^2 - r_w^2\right)$ in Eq. (9.8) changes to R_e^2. The ratio of the vertical flow to well discharge of the well (say n) can be written as

$$n = \frac{\pi R_e^2 v}{Q} \tag{9.10}$$

Incorporating the value of v from Eq. (9.10) in Eq. (9.8), the resultant equation for flow through well (after neglecting r_w with respect to R_e) can be written as

$$Q = \frac{\pi K\left(H^2 - h_w^2\right)}{\log_e \frac{R_e}{r_w} - \frac{n}{2}} \tag{9.11}$$

Incorporating the value of Q from Eq. (9.11) in Eq. (9.10), the resultant equation for the rate of replenishment is given by

$$v = \frac{nK \frac{\left(H^2 - h_w^2\right)}{R_e^2}}{\log_e \frac{R_e}{r_w} - \frac{n}{2}} \tag{9.12}$$

From Eq.s (9.8) and (9.12), the equation for the drawdown curve after neglecting the term $\frac{r_w^2}{R_e^2}$ can be written as

$$h^2 = \frac{\left(H^2 - h_w^2\right)\left\{\log_e \frac{r}{r_w} - \frac{n}{2}\left(\frac{r}{R_e}\right)^2\right\}}{\log_e \frac{R_e}{r_w} - \frac{n}{2}} + h_w^2 \tag{9.13}$$

The Eqs. (9.11) and (9.13) can assess the water level in the area of influence of drainage well.

10

Surface Drainage

10.1 Introduction

Surface drainage may be defined as the diversion or orderly removal of excess water from the land surface by means of improved natural or constructed channels, supplemented when necessary by shaping and grading of the land surface to such channels (ICID, 1982). Usually, surface drainage is applied on flat lands where soils have low permeability, or have less infiltration rate, or having impermeable layer at shallow depth to prevent the ready absorption of high-intensity rainfall. Therefore, the presence of excess water on land surface may be due to the following reasons.

- The permeability of the soils is very low.
- The presence of impermeable barrier at shallow depth (may be due to rock, clay, or lime layer).
- Excessively flat topography with surface depressions where water is accumulated.
- Lack of natural drainage system.

The design criteria of surface drainage system are based on (i) sensitivity of crops to ponded water and saturated soils, and (ii) the engineering considerations of flow through constructed channels and associated structures.

10.2 Negative effects of poor surface drainage

- Poor surface drainage leads to inundation of crops. As a result, the growth of the crop severely hampered and production decreases drastically.
- In temperate regions, soil temperature (required for seed germination) does not rise to required degree due high moisture conditions.
- Saturated conditions lead to removal of oxygen from the root zone, which hampers germination and uptake of nutrients.
- High moisture level in the soil restricts the movement of agricultural machinery for performing mechanized farming operations in every portion of land (in lack of appropriate traction).

10.3 Types of surface drainage systems

Various types of surface drainage systems can be classified as

- Bedding system
- Land grading and land planning
- Random field drainage system
- Parallel field drainage system
- Cross-slope drainage system
- Interceptor or diversion drains

The choice of a particular system depends upon the (i) soil type, (ii) topography, (iii) crops to be grown, and (iv) land owner's preference.

10.3.1 *Bedding system* – Under the bedding system, the beds are separated by parallel dead furrows oriented in the direction of the greatest land slope (Figure 10.1). The dead furrows collect the water drained from the beds (Figure 10.2) and convey to field drain constructed perpendicularly to the dead furrows at the lower end of the field. The major drawback of the system is that the cropped portion adjacent to the dead furrows does not drain satisfactorily. The bed width depends upon the (i) the size of the implements used for various operations in the field, (ii) the permeability and infiltration rate of soil, and (iii) crops to be grown (either field or row crops). To achieve satisfactory performance, the bed width should not be more than 10 m.

Disadvantages of the bedding system

- During the bed formation, the top fertile soil of the sides of bed is moved to the middle for creating slope. It may cause reduction in yield at the bedsides.
- To prevent weed growth in the dead furrows, regular maintenance is required.
- A small mistake in slope while forming dead furrows may lead to water stagnation.
- Mechanized farming is restricted by the bedding system.

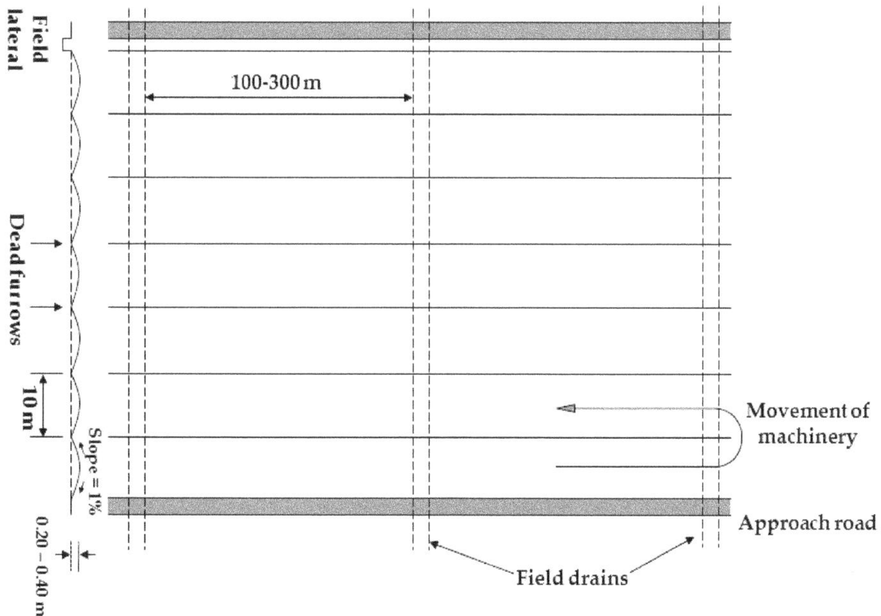

Figure 10.1: Schematic sketch showing the bedding system.

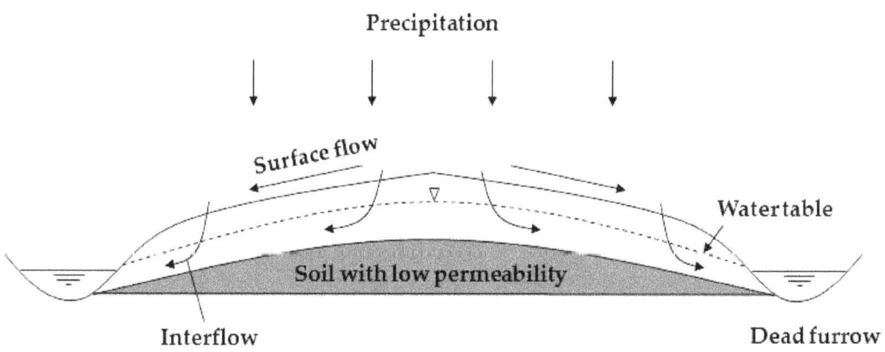

Figure 10.2: Schematic sketch showing the surface and interflow through bed in bedding system.

10.3.2 *Land grading and land planning* – For overcoming the disadvantages of bedding system, ICID (1982) has developed the mentioned methods. In the land grading method, land surface is provided with a predetermined grade to a field drain. It is a one-time operation and while doing so, cutting, filling, and smoothing operations are performed. This method promotes the efficient draining of water along with the mechanized farming. The absence of any field drain in the graded land eliminates

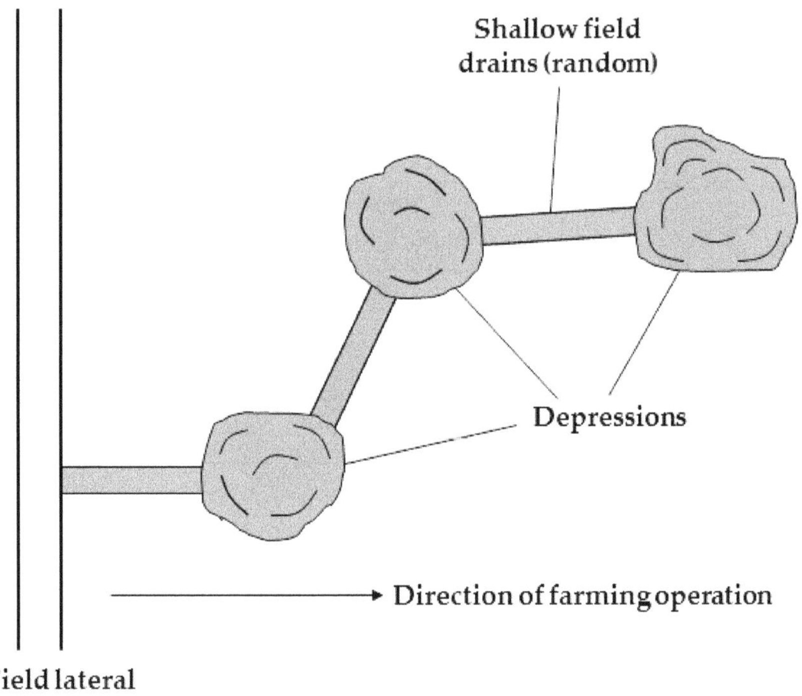

Figure 10.3: Schematic sketch showing the random field drainage system.

the need for maintenance and weed control. Conventional earthmoving equipment in combination with specially designed machinery is used for land grading (Haynes, 1966). Sometimes, the land is in required grade but have minor depressions. Therefore, in the land planning method a land-plane is used to eliminate the minor depressions and other irregularities without changing the grade.

10.3.3 *Random field drainage system* – This system is adapted for draining a field having a number of depressions that are randomly distributed over an area (Figure 10.3). Generally, these depressions are shallow in depth but large in size. Moreover, land smoothening is not economical for these depressions. Therefore, water from these depressions can be removed by constructing random field drains connecting each other and ultimately fall into the lateral. Field drains are constructed with a design so that farm equipment can cross these. Obviously, field drains should be shallow and have flat side slopes. Generally, V-shape is preferred with side slopes not steeper than 6H:1V.

10.3.4 *Parallel field drainage system* – This system is adapted in the flat areas where topography is irregular at micro level. The area is provided with parallel field drains. However, it is not necessary that the drains be equidistant (Figure 10.4). When surface drains are used in conjunction with subsurface drains, the combined system is known as '*Parallel Open Ditch System*'. In the combined system, ditches are laid at steeper slope and are deeper than surface drains. The farming operations

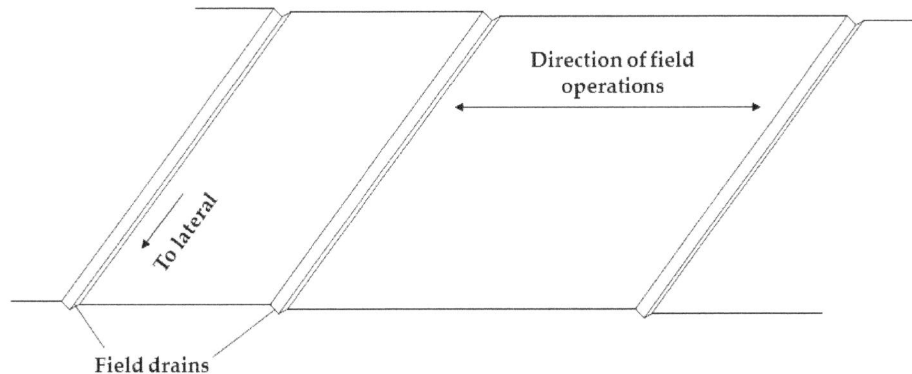

Figure 10.4: Schematic sketch showing the parallel field drainage system.

are performed parallel to these ditches, as farm equipment cannot be cross these. The combined system of surface and subsurface drains is usually adapted for peat and muck soils (Ritzema, 1994).

10.3.5 *Cross-slope drainage system* – It is a channel-type graded terrace and is also known as *Nichlos terrace*. This system is adapted in the areas having slope of up to 4%. As far as possible, the drains should be formed along the contours of the land at a grade ranging from 0.1 to 1.0%. The graded area between the drains is smoothened for farming operations. All the farming operations are performed parallel to the drains. Usually, triangular or trapezoidal cross-section is preferred for constructing cross-slope drains with side slope ranging from 4H:1V to 10H:1V. The general dimensions of the drain are depicted on the Figure 10.5. It is easy to maintain a cross-slope drain of length ranging from 350 to 450 m. The distance between the drains is kept in the range of 30 to 45 m and the factors like land slope, rainfall intensity, soil erodibility, and crops to be grown affect this value.

10.3.6 *Interceptor or diversion drain* – To protect flat lands from flooding by surface runoff from adjacent sloppy grounds, an interceptor or a diversion drain is constructed at the foot of these uplands. A drain with a minimum depth of 0.45 m and with a minimum cross-sectional area of about 0.70 m^2 is considered. The interception system is also known as cross-slope ditch system and is adapted in the regions having slope less than or equal to 4%. In general, the spacing between the adjoining ditches is kept equal to 30 m on a 4% slope. Ditch spacing increases to 50 m if general slope of land decreases to 0.5%. After making the interception ditches, farming operations are performed parallel to the ditches.

10.4 Open ditches

For surface and subsurface drainage, generally open ditches (as individual or main drain) are used. Usually, these ditches act as collectors and carry the flow received from the subsurface drainage system. The various advantages of the open ditches are

- Their construction is very easy and therefore, no specialized machines are required.

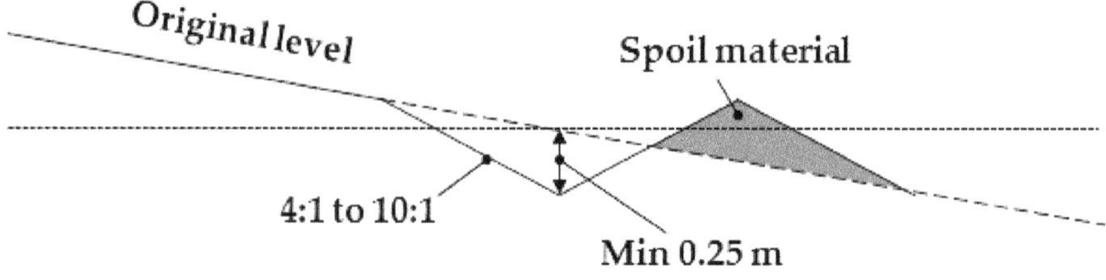

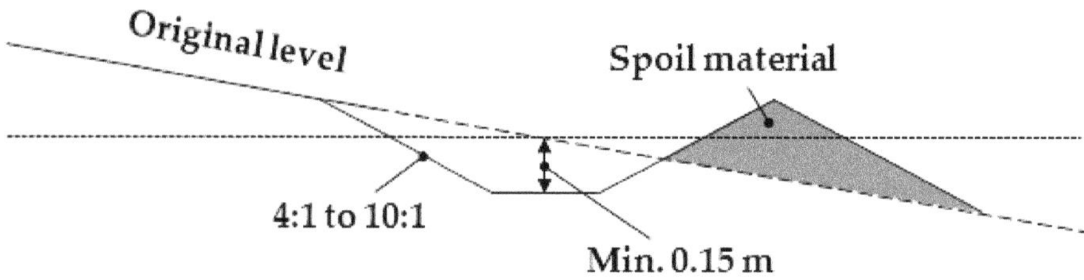

Figure 10.5: Schematic sketches showing the parallel field drainage system.

- It is convenient to dig the drain manually (depending upon the availability of labor economically).
- Open ditches can carry large quantities of water because of their large size.
- It is most convenient way of handling the runoff caused by precipitation.
- These can also be employed to control water table.

There are few disadvantages of the open ditches, which are written below

- Their presence in the cultivable lands obstructs the farming operations, as farm machinery cannot cross the ditches. There may be considerable loss in production.
- It is very difficult to maintain the open ditches. With time, weeds and wild plants grow in the drains, which obstruct the movement of water through it.
- If the cost of the land is very high, the system becomes costly. Moreover, for crossing these drains bridges are required, which further increases the cost.
- It is difficult to keep the system under operational conditions in unstable soil conditions.

10.5 Design of open ditches

10.5.1 *Channel cross section* – The most efficient channel cross section is one having with the smallest wetted perimeter. It can be properly explained using the manning's formula which can be written as

$$Q = AV = A \cdot \frac{R^{2/3} S^{1/2}}{n} = A \cdot \frac{\left(\frac{A}{P}\right)^{2/3} S^{1/2}}{n} \tag{10.1}$$

In Eq. (10.1), R is the hydraulic radius and can be evaluated by dividing wetted cross-sectional area (A) of the channel by its wetted perimeter (P). It is obvious that with the increase in P, Q decreases. Semi-circular channel section has maximum hydraulic radius. However, trapezoidal channel cross-section is usually preferred for ease in its construction and for high hydraulic radius.

10.5.2 *Side slopes* – Field experience shows that for clayey and highly organic soils, a side slope of 1H: 1V and for sandy soils, a side slope of 3H: 1V remains stable.

10.5.3 *Scouring and deposition in ditch* – The scouring and deposition in ditch depend upon the soil type and its grade. At the steeper reaches in ditch, erosion takes place, and the erosive material is deposited at flatter portions. While fixing the grade of ditch for a particular soil type, government agencies recommended practices should be followed.

10.5.4 *Character of spoil* – The soil removed during the construction of ditch is usually piled some distance away from the side of ditch (Figure 10.6). The placement of spoil

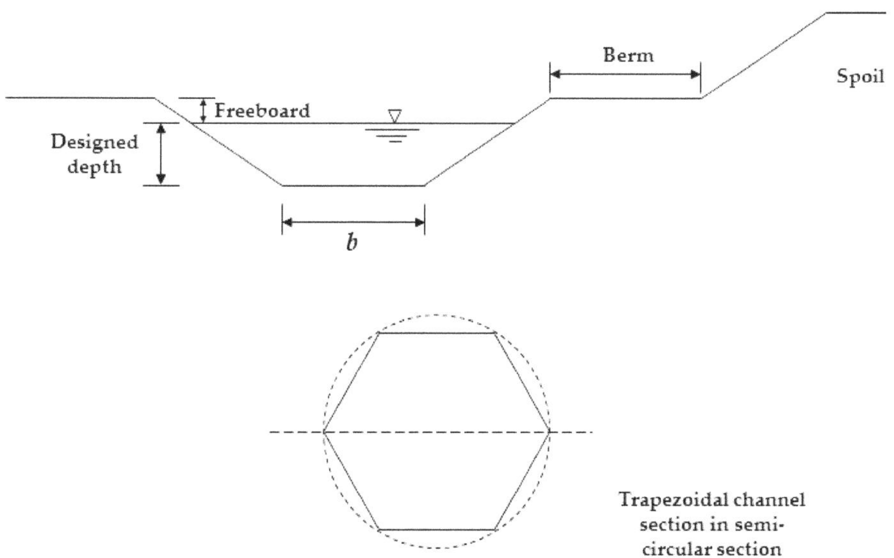

Figure 10.6: Schematic sketch showing the elements of a ditch for drainage purpose.

adjacent to the side of the ditch may reduce its stability due to additional weight. The area lies between the side of ditch and spoil bank is termed as *berm*. Sometimes, the *berm* is used as pathway for moving machinery.

10.6 Maintenance of open ditches

Open ditches should be maintained on yearly-basis for keeping these in operative conditions.

10.6.1 *Entry of sediments in an open ditch* – Erosion of the surrounding lands generates the sediments, which enter the open ditches. The adaption of erosion control structures can help in reducing the dislodgement of sediments. The scouring at some place and deposition of sediments at another place is also possible in the ditch. To promote the settling of sediments (specifically silt) at one location for ease in cleaning, sediment basins or low check dams are provided in the ditch.

10.6.2 *Protection of side slope of ditch* – By planting local grasses, side slope of the ditch can be protected against cave-in.

10.6.3 *Control of bank-erosion* – By providing obstructions (i.e. jetties) in the path of water striking the particular portion of the bank, erosion can be controlled. Brush revetment or local rocks are used for the purpose.

10.6.4 *Weed control* – Growth of weeds in the ditch reduces its water carrying capacity. Weeds can be controlled by chemical/mechanical means. By dragging a chain in the ditch, the weeds are uprooted. High temperature burning of weeds is another option.

10.7 Methods of constructing ditches

For digging smaller depth ditches or for giving finishing touch to the machine dug ditches, *manual labor* is used. According to the conditions of land, different machines are used for digging ditches. These are classified in Table 10.1.

Table 10.1: Classification of ditch digging machines.

S. No.	Machine name	Purpose
1	Drain plow	For constructing drainage ditches in low value lands
2	Clamshell	For digging straight down, sumps, or trenches that are sheathed or have crisscross bracing. These work excellently underwater.
3	Dragline	For excavating loose materials below the grade of the machine. These can dig beneath the water table.
4	Backhoe/trench hoe/drag shovel	For digging below the ground level very precisely.

10.8 Mole drainage system

Heavy soils of low hydraulic conductivity (less than 0.01 m/d) often require very closely spaced drainage systems (2 – 4 m spacing) for satisfactory water control. Mole drains are unlined circular soil channels which function as pipe drains. The moles can be formed at very low cost, and hence the system can be installed

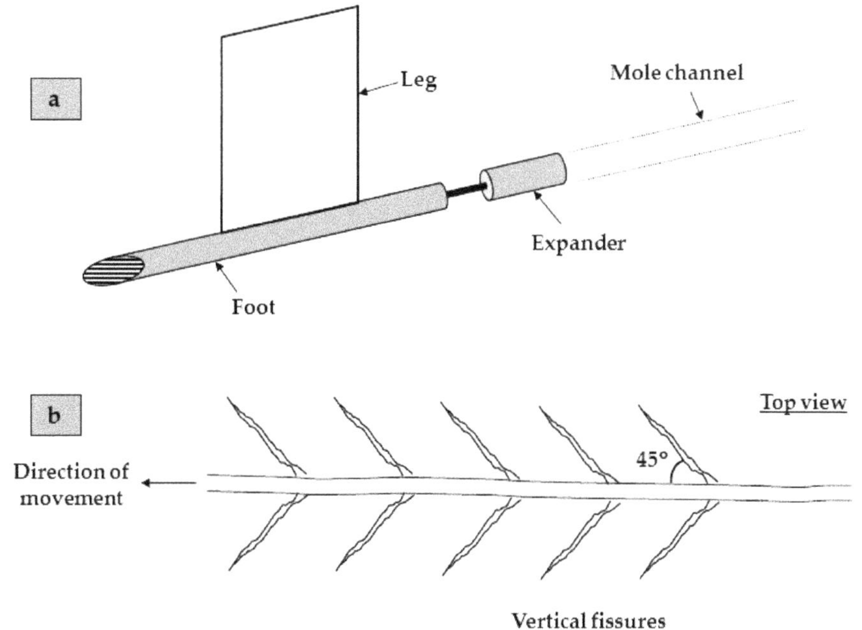

Figure 10.7: Schematic sketches showing the (a) mole plough, and (b) development of vertical fissures in the soil due to the movement of mole plough.

economically at very close spacing. However, the service life of the system is not very long. Mole drains are formed with mole plough, which comprises a cylindrical foot attached to a narrow leg, followed by a slightly larger diameter cylindrical expander (Figure 10.7a).

The foot and the expander form the drainage channel and the leg generates a slot with associated soil fissures which extend from the surface down into the channel. The leg fissures are vertical and are formed at an angle of approximately 45° to the direction of travel (Figure 10.7b). Commonly used mole plough has the following dimensions.

- Foot diameter = 7.5 cm
- Expander diameter = 10 cm
- Leg thickness = 2.5 cm
- Side length of leg = 20 cm

The number and size of the leg fissures produced with a given mole plough are dependent upon soil conditions. A smaller number of wide fissures tend to form under drier conditions, but as the soil-water conditions become increasingly plastic, the fissures become narrower and more numerous. Mole channel walls become smoother as soil-water content increases. The following expander increases the smoothing effect.

11
Salt Affected Soils

11.1 Introduction

While applying irrigation, we are also inserting salts in the soil. Even good quality water contains soluble salts. With proper leaching or irrigation practices, these salts leached down to water table. To avoid the capillary rise of salts, this water must be drained off. In other terms, irrigation is balanced by the drainage (either natural or artificial). Therefore, in irrigated soils, the aim of drainage is to control the soil salinity; whereas, in humid regions, its motive is to control soil water for better aeration, higher temperature, and easier workability.

Indiscriminate use of available water as irrigation without studying the dynamics of groundwater in the region may cause waterlogging. Consequently, the accumulated salts at the surface soil leads to the unfavorable soil-water-plant relationships. Gradually, the region goes out of cultivation. In different regions of India, salt affected soils are designated with different names such as *Kallar, Thar, Reh, Rakkar*, and *Kulrati* in Punjab, Haryana, U.P., Uttarakhand, and Rajasthan; *Luni, Karl*, and *Chopan* in Maharashtra; *Palachoudu, Choudu, Uippu*, and *Karu* in Tamil Nadu; *Pokhali, Kari*, or *Kaipad* in Kerala; and *Phodus* in Telengana and Andhra Pradesh.

11.2 Definitions

Cation exchange capacity (*CEC*) – The total quantity of cations that a soil (with active surface) can adsorb by cation exchange is known as CEC. Its units are milliequivalents/100g.

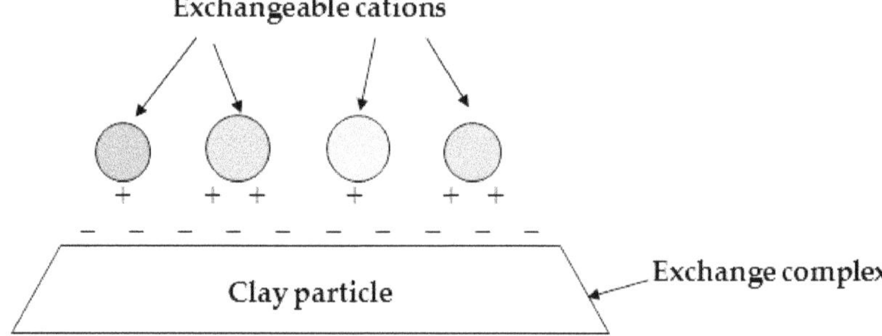

Figure 11.1: Schematic sketch showing the exchange complex and exchangeable cations.

Milliequivalent (*meq*) – A milliequivalent is the mass of an ion or compound that combines with or replaces 1 mg of hydrogen and can be evaluated by dividing atomic mass of the ion by its valency.

Exchange complex – The soil's surface-active constituents those have ability of cation exchange is known as exchange complex (Figure 11.1).

Exchangeable cation – A cation on the surface-active material (i.e. soil or exchange complex) and which can be exchanged with another cation is known as exchangeable cation (Figure 11.1).

Diffuse double layer (*DDL*) – The diffuse accumulation zone of cations is known as DDL. In the DDL, cations are attracted by the surface of solid phase and at the same time, anions are excluded.

Exchangeable sodium percentage (*ESP*) – The degree of saturation of the exchange complex with sodium is known as ESP and can be calculated as

$$ESP = \frac{Exchangeable\ Sodium}{CEC} \times 100 \qquad (11.1)$$

Sodium adsorption ratio (*SAR*) – SAR is the ratio to express the relative activity of sodium ions in the soil and in the irrigation water. Mathematically, it can be written as

$$SAR = \frac{Na^{+}}{\sqrt{\frac{\left(Ca^{2+} + Mg^{2+}\right)}{2}}} \qquad (11.2)$$

Electrical conductivity (*EC*) – A straight-line relationship occurs between the EC of a soil extract and the soil's salt concentration when plotted on a graph sheet. Therefore, the salt content of a soil is commonly expressed by EC. It is measured at a reference temperature of 25°C. The reciprocal of electrical resistivity is known as EC. The formula for resistance can be written as

$$R = \rho \frac{L}{A} \tag{11.3}$$

In Eq. (11.3), R is resistance (ohm); ρ is the electrical resistivity (ohm-cm); L is the length of conductor (cm); and A is the cross-sectional area of the conductor (cm^2). The units of EC will be the reciprocal of the units of electrical resistivity (i.e. $(ohm-cm)^{-1}$ or $\frac{mho}{cm}$). The smaller unit of EC (i.e. $\frac{mmho}{cm}$) can also be written as $\frac{dS}{m}$.

pH – The acid or base character of any (aqueous) solution can be defined by means of single variable, i.e. the hydrogen ion activity. Numerically, the pH of a solution is the logarithmic of the number of liters of the solution that contain one-gram atomic weight (1.008 g) of hydrogen as (H^+) ions. Mathematically, it can be written as

$$pH = \log_{10}\left[\frac{1}{H^+}\right] \tag{11.4}$$

The pH of the neutral water is 7.0. The presence of calcium and magnesium carbonates in water is reflected by the pH value of 7.5 to 8.0. The pH value of 8.5 or more indicates the presence of exchangeable sodium.

Saturation extract – The solution extracted from a soil by filtering its water saturated solution.

Dispersed soil – The presence of excess sodium in soil disperses it. In lack of definite soil structure, the permeability of such soils reduces considerably. These soils become plastic or slurry on wetting and become hard on drying.

Specific surface – The ratio of surface area of soil to its unit weight is known as specific surface. Usually, its units of measurements are m^2/g.

Osmotic pressure – It is the equivalent negative pressure, which influences the rate of diffusion of water through a semi-permeable membrane.

11.3 Types of soils

11.3.1 *Alkali soils* – These are also known as *sodic soils*. These soils, in general, contain sodium carbonate (Na_2CO_3), sodium bicarbonate ($NaHCO_3$), and sodium silicate (Na_2SiO_3). The presence of sufficient exchangeable sodium (imparted from the mentioned salts) deteriorates the physical conditions of soils and thus affects the crop production. A soil having ESP above 15% and EC of the saturation extract less than 4 dS/m (at 25°C) is an alkali soil. As mentioned earlier that pH value of alkali soils is more than 8.5 and ranges from 8.5 to 10.0.

11.3.2 *Saline soils* – In saline soils, chlorides and sulphates of sodium, calcium, and magnesium. Nitrates may be present in appreciable quantities only rarely. Soluble carbonates are always absent. A soil having ESP less than 15% and EC of the saturation extract more than 4 dS/m (at 25°C) is a saline soil. Moreover, the pH of the saline soils is generally less than 8.5. Saline soils are rich in neutral salts. The normal growth of plants is affected by (i) high osmotic pressure, and (ii) toxic effect of individual ions present in the soil solution.

11.3.3 *Saline-alkali soils* – A soil having ESP above 15% and EC of the saturation extract more than 4 dS/m (at 25°C) is a saline-alkali soil. Usually, the pH of the saline-alkali soils remains less than 8.5 and the sodium ions are present as neutral salts as sodium chloride ($NaCl$) and sodium sulphate (Na_2SO_4). In case, the pH of the soil is more than 8.5, then it indicates the presence of bicarbonate (HCO_3^-) and carbonate (CO_3^{2-}) ions.

11.4 Relation of EC with the salt/total-cation concentration and osmotic pressure

Salt concentration (mg/l) = 640 × EC (dS/m) (11.5)

Total cation concentration (meq/l) = 10 × EC (dS/m) (11.6)

The Eq. (11.6) is valid for the condition that EC should not be more than 5 dS/m at 25°C.

Osmotic pressure (atm) = 0.36 × EC (dS/m) (11.7)

11.5 Causes of salt build-up in irrigated soils

Mostly, salt affected soils occur in arid and semi-arid regions. At present, approximately 6.75 million hectares of land area is salt affected and it is projected that in year 2025, this figure will rise to around 12 million hectares. The major causes for its development are

- Rising water table due to indiscriminate use of available water. There are regions all over the world where the quality of native groundwater is brackish. At present, around 25% of the total water used for irrigation is either saline or brackish. Non-withdrawal of this native brackish water and excess irrigation rate is a cause of rising water table. Seepage from canals is also responsible for rising water table. In high water table regions, salts move up under thermal gradient and is accumulated on the land surface.
- Inadequate natural drainage outlets of an area.
- Irrigation with poor quality water.
- Sea water intrusion (restricted to coastal regions only).

In general, light textured soils are less salinized than the medium and heavy textured soils. In water shortage regions, poor quality water is used to irrigate the crops. While irrigating with poor quality water, the various processes occurring in the soil are as follows

- The $CaCO_3$ present in the soil dissolute and precipitate.
- Hydration and dehydration of soils.
- Ions leach down with percolated water and may move upward through capillary activity.
- Cation exchange between irrigation water and soil.

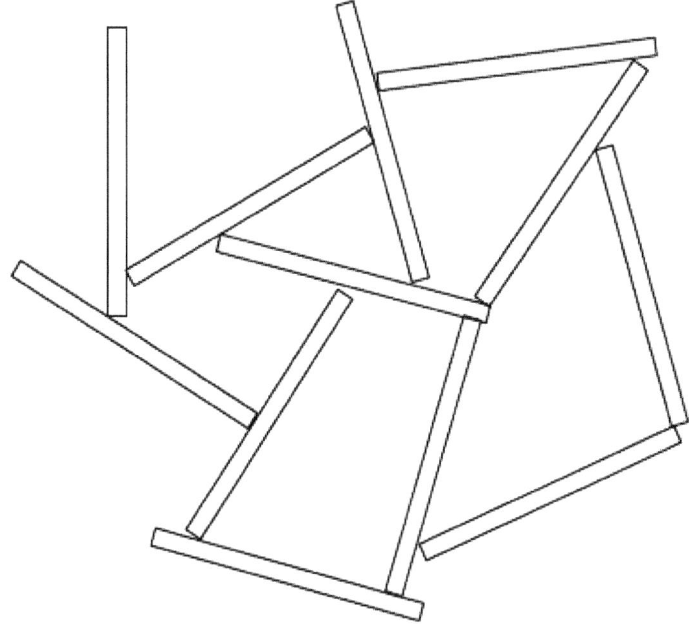

Figure 11.2: Clay particles in card-house arrangement.

11.6 Sodium effect on soil physical behavior

It is a well-known fact that negative charge on the soil exchange complex is counterbalanced by an equivalent amount of cations present in the adjacent liquid phase. Diffusion phenomenon distributes the cations evenly throughout the liquid phase. Ultimately, the cations attracted towards the surface of the solid phase by the electric field. The divalent cations (like Ca^{2+}) attracted to the exchange complex with more force in comparison to monovalent cations (like Na^{+}). If Ca/Na ratio is increasing in the soil-water system, the extent of the DDL will decrease and vice-versa. Soil's physical behavior is greatly affected by the extent of DDL. If the extent of the double layer is small, then a clay particle has a tendency to attract neighboring clay particle to form weakly bonded floccules. The clay particles form a card-house arrangement (Figure 11.2). The presence of organic matter, lime, and gypsum favors the stability of these floccules (Ritzema, 1994). At field capacity, the DDL is developed up to its potential extent. In other words, the DDL swells on wetting and shrinks on drying. However, this swelling is far less for soils high in exchangeable calcium than for soils high in exchangeable sodium. Obviously, the swelling of the DDL causes the closure of the inter-aggregate pores. Consequently, hydraulic conductivity of the soil reduces drastically. The increase in the pressure due to swelling forces the clay particles away from each other. As a result, soil dispersion takes place. The detached fine soil particles clog the soil pores, which further reduces soil permeability.

We are considering that ESP more than 15% is hazardous for soil structure. In fact, this limit changes with the soil type. For examples, for sandy soils the ESP up to 25% may not be harmful; whereas, for clayey soils, only 5% ESP is detrimental for soil structure (Ritzema, 1994).

11.7 Exchange process

Clay is the finest fraction of soil, which has certain charge deficit. This charge deficit is balanced by the positively charged cations (replaceable) present in the irrigation water. Different ions have different tendencies to replace or be replaced by other ions and this order can be written as

$$NH_4^+ > Mg^{++} > Ca^{++} > K^+ > Na^+ \quad (11.8)$$

The replace-ability of Na^+ by Ca^{++} is explained in the following reaction

$$^{Na}_{Ca}\langle Clay|\,particle\,\rangle^{Na}_{Na} + CaCO_3 \Leftrightarrow Na_2CO_3 + {}^{Na}_{Ca}\langle Clay|\,particle\,\rangle^{Ca} \quad (11.9)$$

The salt Na_2CO_3 is soluble in water and leaches down.

11.8 Salt balance

The equation for salt balance can be written as

$$V_{iw}C_{iw} - V_{dw}C_{dw} = S_m + S_p + S_c \quad (11.10)$$

In Eq. (11.10), V_{iw} and V_{dw} are the volume of irrigation and drainage water, respectively; C_{iw} and C_{dw} are the salt concentrations in irrigation and drainage water, respectively; S_m is the salt dissolved from soil minerals; S_p is the amount of salt precipitated over the soil surface; and S_c is the amount of salt removed by the crop. In comparison to other major components, the terms S_m, S_p, and S_c are very small and can be neglected. Therefore, the Eq. (11.10) changes to

$$V_{iw}C_{iw} = V_{dw}C_{dw} \quad (11.11)$$

The terms V_{iw} and V_{dw} can be written as

$$V_{iw} = Area\ under\ drainage\ network \times D_{iw} \quad (11.12)$$

$$V_{dw} = Area\ under\ drainage\ network \times D_{dw} \quad (11.13)$$

Incorporating Eq.s (11.12) and (11.13) in Eq. (11.11), it changes to

$$\frac{C_{iw}}{C_{dw}} = \frac{D_{dw}}{D_{iw}} \quad (11.14)$$

11.9 Leaching requirement

Leaching requirement is that fraction of irrigation water, which must be leached through the root zone of plants in order to prevent the soil salinity from exceeding a specified level. It depends upon the depth of irrigation water as well as on the depth of drainage water. As per definition, it can be written as

$$LR = \frac{D_{dw}}{D_{iw}} \tag{11.15}$$

Since, salinity of water is measured in terms of EC, the Eq. (11.15) can also be written as

$$LR = \frac{D_{dw}}{D_{iw}} = \frac{EC_{iw}}{EC_{dw}} \tag{11.16}$$

In saline soils, depth of irrigation water is the sum of drainage water and consumptive use (D_{dw}) as

$$D_{iw} = D_{cw} + D_{dw} \tag{11.17}$$

Combining Eq.s (11.16) and (11.17) by eliminating the term D_{dw} the resultant equation can be written as

$$D_{iw} = \left(\frac{EC_{dw}}{EC_{dw} - EC_{iw}} \right) D_{cw} \tag{11.18}$$

11.10 Reclamation of salt-affected soils

Before starting the reclamation of salt-affected soils, preliminary information in detail is required. The first query is regarding the basic cause of soil salinization, which may be due to shallow water table, irrigation with poor quality water or seawater intrusion. Whether the soil contains soluble salts or there is any sodium hazard? If so, then what is the exchangeable sodium percentage? Present physical characteristics of soil should be known. Moreover, the hydraulic conductivity of the soil should be known at three places (i.e. top, subsoil, and the substratum). For leaching of salts present in the soil, availability of a perfect drainage system and freshwater are the prime requirements.

11.10.1 *Leaching methods for saline soils* – Ponding a required depth of water on the land surface is the effective way of creating downward movement of water. On flat lands, rectangular checks and level borders are used, whereas contour checks are preferred for sloping lands. However, leaching by furrows is limited. Leaching methods can further be divided into three categories *viz.* (i) continuous ponding, (ii) intermittent ponding, (iii) sprinkling. For leaching saline soils, water is continuously ponded on the field for long periods and is most common practice. In intermittent leaching method, water is applied at intervals of one to two weeks. During the interval, soil becomes dry. We know that cracks are formed in clayey soils on drying. With the formation of cracks, the evaporative surface area of the soil increases and salts are accumulated on the surface, which are washed out during the next application of water. The most important point, which differentiates the intermittent ponding and sprinkling method from continuous ponding, is the considerable saving of water. However, with intermittent method, more time is required for reclaiming the soil. In addition, for sprinkling method, specialized equipment is required.

Another important factor during the leaching of salts is the *uniformity of the salt removal*. During the ponding conditions, the complete soil profile becomes saturated. In chapter 7, the movement of water towards the drains has been shown. Therefore, inference can be drawn that the salt displacement from the area midway between the drains will be less than the areas close to the drains. *Surface flushing* is another technique, which is used to remove visible crusts of salt on the surface of salt-affected soils. Nonetheless, the method is not very effective.

11.10.2 *Reclamation of alkali soils* – Usually, the alkali soils are without any soil structure. Therefore, leaching is practically not feasible in this case until a proper soil structure has not been formed. Therefore, chemical treatment is performed before adapting leaching operation. These chemicals act in two-ways. Either these contain soluble calcium or they enhance the solubility of calcium carbonate ($CaCO_3$) if present in the soil. Organic matter, Sulphur, pyrites, or acid waste products such as press-muds and waste sulphuric acid enhance the solubility of calcium carbonate and therefore, are restricted to soils containing $CaCO_3$ only (Ritzema, 1994). Calcium chloride ($CaCl_2$), gypsum ($CaSO_4 \cdot 2H_2O$), and phosphogypsum($CaSO_4 \cdot 2H_2O$) are the chemicals which directly add the calcium in soil. It is worth to mention here that the phosphogypsum refers to the gypsum formed as a by-product of the production of fertilizer from phosphate rock. It contains around 80-90% calcium sulphate and is very effective in reclaiming alkali soils (Mehta, 1983). The efficiency of the gypsum application depends on the factors like (i) non-uniform spreading of gypsum, (ii) other than sodium, calcium from the gypsum also replaces the magnesium and potassium in the soil, (iii) part of the gypsum escapes through drainage without reacting with soil, and (iv) fineness of the gypsum salt.

11.10.3 *Reclamation of saline-alkali soils* – Structure of saline-alkali soils is in superior condition than the alkali soil. However, the application of water for leaching of salts may breakdown the soil structure making the reclamation tedious and time consuming. Just like the reclamation of alkali soils, chemical amendments are applied (Abrol et al., 1988) before leaching of the end products.

Objective Type Questions Based Upon the Chapters 1 to 4

Fill in the blanks.

1. The formations within the saturated zone below the ground surface from which water can be tapped are called ________.

   a. Aquifer b. Water bearing formation
   c. Aquitard d. Both (a) and (b)

2. The water level in wells penetrating________ aquifer does not rise above the phreatic level in general.

   a. Confined b. Unconfined
   c. Semi-confined d. Both (b) and (c)

3. __________ refers to the upper surface of the zone of saturation.

   a. Water table b. Phreatic surface
   c. Piezometric surface d. Both (a) and (b)

4. A confined aquifer is also called as________ aquifer.

   a. Pressure b. Artesian
   c. Both (a) and (b) d. None of above

5. An aquifer is bounded by an impermeable layer at the bottom and by a semi-pervious layer above it. It is termed as ___________.

   a. Semi-pervious aquifer b. Leaky aquifer
   c. Confined aquifer d. Both (a) and (b)

6. The water level in a well penetrating a confined aquifer represents the ________ in the aquifer.

   a. Artesian pressure b. Phreatic pressure
   c. Tension d. None of above

7. The elevation to which water level will rise in a well tapping a confined aquifer is called as________.

    a. Pressure level
    b. Piezometric level
    c. Isopiestic surface
    d. Both (b) and (c)

8. Piezometric surface in ______ is analogous to water table in an unconfined aquifer.

    a. Confined aquifer
    b. Semi-confined aquifer
    c. Bounded aquifer
    d. Both (a) and (c)

9. When the piezometric surface of a completely confined aquifer lies above the land surface, a well installed in this aquifer at the place may be termed as ________.

    a. Artesian well
    b. Flowing well
    c. Both (a) and (b)
    d. None of the above

10. A_________ layer refers to a formation having low but measurable hydraulic conductivity.

    a. Aquitard
    b. Semi-pervious
    c. Leaky
    d. All above

11. While pumping a ____________, both phreatic and piezometric levels are lowered simultaneously.

    a. Leaky aquifer with prompt yield
    b. Leaky aquifer with delayed yield
    c. Perched aquifer
    d. Confined aquifer

12. While pumping a ____________, there is a considerable lag between the lowering of phreatic and piezometric levels.

    a. Leaky aquifer with prompt yield
    b. Leaky aquifer with delayed yield
    c. Perched aquifer
    d. Confined aquifer

13. The wells can be used for recharging groundwater if ____________.

    a. Land is not available for water spreading
    b. Hard pan or impervious layer at shallow depth
    c. Targeted aquifer is below an unconfined aquifer
    d. All above

14. The geophysical methods provide ________ evidence of the subsurface formations.

    a. Indirect b. Direct
    c. Quantitative d. Qualitative

15. Electrical resistivity of aquifer formation containing ________ show high resistivity.

    a. Gravel b. Clay and shale
    c. Sand and sandstone d. Both (a) and (c)

16. In electrical resistivity method, clean sand saturated with freshwater shows ________ resistivity value.

    a. Low b. High
    c. Zero d. Infinite

17. Electrical resistivity of the underground formations can be measured only in ____________.

    a. Cased boreholes b. Mud filled holes
    c. Uncased boreholes d. Both (b) and (c)

18. Electrical resistivity surveying is affected by ____________.

    a. Potential drop b. Applied current
    c. Electrode spacing d. All above

19. In gamma ray logging method, the emission of gamma rays in counts per second is plotted against ________.

    a. Time b. Depth
    c. Horizontal distance from bore well d. Either (a) or (c)

20. In electrical resistivity surveying, the ratio of the potential drop to the applied current is referred to as the apparent resistivity of the lithosphere down to a depth, which is proportional to __________.

    a. Electrode's insertion depth b. Thickness of the formation
    c. Electrode's spacing d. Groundwater potential

21. Seismic refraction surveying is suitable only where ____________.

a. Velocity of the shock wave increases with depth
b. Velocity of the shock wave decreases with depth
c. Velocity of the shock wave is not affected by the formation thickness
d. Velocity of the shock wave remains constant with depth

22. Hydraulic conductivity is a coefficient, which depends on ___________.

a. Formation properties
b. Fluid properties
c. Intrinsic permeability of formation material
d. Both (a) and (b)

23. In general, the transmissivity of an aquifer ranges between ______________.

a. 12 – 1200 m^3/d/m
b. 12 – 120 m^3/d/m
c. 12000 – 1200000 l/d/m
d. Either (a) or (c)

24. The term coefficient of storage is related to _______.

a. Confined aquifers
b. Unconfined aquifers
c. Semi-confined aquifers
d. Both (a) and (c)

25. In most confined aquifers, the values of storage coefficient fall in the range __________.

a. 10^{-5} to 10^{-3}
b. 10^{-4} to 10^{-3}
c. 10^{-3} to 10^{-2}
d. 10^{-2} to 10^{-1}

26. The hydraulic resistance of a leaky layer is evaluated using the formula $c = \frac{b'}{K'}$. In this K' represents the ________.

a. Horizontal hydraulic conductivity of leaky layer
b. Horizontal hydraulic conductivity of aquifer
c. Vertical hydraulic conductivity of leaky layer
d. Vertical hydraulic conductivity of aquifer

27. If hydraulic resistance of a geological formation is near zero, it may be treated as _________.

a. Aquifer
b. Aquiclude
c. Aquitard
d. Aquifuge

28. The leakage factor is a term related to _________.

a. Leaky aquifer
b. Leaky layer
c. Both (a) and (b)
d. Aquitard

29. A b' thick semi-pervious layer having hydraulic conductivity K' is overlying an aquifer of thickness b and hydraulic conductivity K. Its leakage factor will be equal to ____________.

a. $B = \sqrt{\frac{Kbb'}{K'}}$

b. $B = \sqrt{\frac{KK'b'}{b}}$

c. $B = \sqrt{\frac{K'b'b}{K}}$

d. $B = \sqrt{\frac{KK'b}{b'}}$

30. In case of flowing wells, static water level lies ___________.

a. Above the water table

b. Above the ground surface

c. At the level of water table

d. Both (a) and (b)

31. An unconfined aquifer is lying above a leaky aquifer. The height to which water will rise in a piezometer installed in the leaky aquifer will tell the ________.

a. Piezometric pressure in leaky aquifer

b. Water level in unconfined aquifer

c. Both (a) and (b)

d. Nothing

32. The well casing of a cavity well ________.

a. Terminates at the upper confining layer

b. Terminates at the bottom of the aquifer

c. Terminates at the top of the aquifer

d. Hangs from the surface

33. While pumping a phreatic aquifer, lowering of phreatic surface beyond a certain distance from the well becomes negligible is showed by _______.

a. Adolph Thiem

b. Theis

c. DeGlee

d. Jacob

34. For small drawdown values, an unconfined aquifer may be treated as ____________.

a. Confined aquifer

b. Semi-confined aquifer

c. Water-table aquifer

d. None of the above

35. Rate of decline of head while pumping when multiplied by storage coefficient and summed over area of influence equals ______________.

a. Discharge rate

b. Discharge

c. Discharge per unit aquifer thickness

d. None of the above

36. An unconfined aquifer is lying above the leaky aquifer. The movement of water through the leaky layer will be proportional to _______.

    a. Difference in phreatic and piezometric heads
    b. Piezometric head above the leaky aquifer
    c. Phreatic level above the unconfined aquifer
    d. None of the above

37. In steady pumping conditions, the quantity of water moving through leaky layer is ________.

    a. Inversely proportional to thickness of leaky layer
    b. Proportional to hydraulic conductivity of leaky layer
    c. Thickness of semi-confined aquifer
    d. Both (a) and (b)

38. For determining steady state drawdown in a ________ aquifer, De-Glee's equation is used.

    a. Semi-confined
    b. Leaky
    c. Un-confined
    d. Both (a) and (b)

39. The criteria required for selecting the pumping test site are ___________.

    a. Distance from railway line, highway, building
    b. Disposal of pumped water
    c. Accessibility
    d. All above

40. Hydraulics of a semi-confined aquifer with delayed yield can be expressed by ___________.

    a. DeGlee's well function
    b. Boulton's well function
    c. Well function
    d. Hantush's well function

41. ________ are required around the pumping well to study distance-drawdown relationship.

    a. Observation wells
    b. Piezometers
    c. Abandoned wells
    d. All above

42. Drawdown distribution curve due to pumping of an aquifer propagates faster in __________.

    a. Unconfined aquifers
    b. Confined aquifers
    c. Semi-confined aquifers
    d. Both (b) and (c)

43. The radius of influence will be greater in ________.

 a. Confined aquifer with high hydraulic conductivity
 b. Unconfined aquifer with high hydraulic conductivity
 c. Confined aquifer with low hydraulic conductivity
 d. Unconfined aquifer with low hydraulic conductivity

44. Theis method of solution for the evaluation of formation constants is based on relation between *W(u)* and *u* which is similar to that ____________.

 a. Between r and s^2/t
 b. Between s and r^2/t
 c. Between u and s
 d. Between u and r

45. It is difficult to fabricate and install an observation well of diameter less than about __________.

 a. 5.0 cm
 b. 7.5 cm
 c. 2.5 cm
 d. 1.5 cm

46. In general, it takes around _____ to reach steady state conditions while pumping unconfined aquifers.

 a. 15-20 h
 b. 24-40 h
 c. 6-8 h
 d. More than 48 h

47. In general, it takes around _____ to reach steady state conditions while pumping confined aquifers.

 a. 24-40 h
 b. 6 h
 c. 18 h
 d. More than 48 h

48. A life span of about ___________ is assumed to design a well.

 a. Less than 10 years
 b. 5 to 7 years
 c. 10 to 15 years
 d. More than 20 years

49. Dykes are natural barriers to groundwater movement and have ___________.

 a. Low porosity and high specific gravity
 b. High porosity low specific gravity
 c. Low porosity and specific gravity
 d. High porosity and specific gravity

50. Open wells are dug down to about 7-10 m below the ____________.

a. Ground surface
b. Static water table in dry season
c. Static water table in monsoon season
d. None of the above

51. To facilitate __________, well curb is placed at the bottom of well.

a. Proper placement of well in the formation
b. Smooth movement of well in the formation
c. Sinking of the well
d. None of the above

52. The well casing diameter should be at least _________ larger than the nominal diameter of the pump.

a. 5 – 10 cm
b. Less than 5 cm
c. 2 – 3 cm
d. As per convenience

53. The well yield is a function of _________.

a. Diameter of casing pipe
b. Diameter of intake pipe
c. Both (a) and (b)
d. None of the above

54. Keeping all other conditions identical, doubling the diameter of a well in unconfined aquifer will increase the yield by about _______.

a. 10%
b. 1.5 times
c. 15%
d. 25%

55. Keeping all other conditions identical, doubling the diameter of a well in confined aquifer will increase the yield by about _______.

a. 7%
b. Two times
c. 10%
d. 25%

56. Considering the same discharge rate, the cone of depression _______ for wells in confined aquifer as compared to unconfined aquifer.

a. Extends to smaller radial distance
b. Extends to larger radial distance
c. Remains same
d. None of the above

57. ________ of the pipe used for tube wells must have a capacity to handle the weight of the assembly plus the frictional force caused by earth pressure.

    a. Tensile strength
    b. Compression strength
    c. Crushing strength
    d. Shearing strength

58. While installing well in a homogeneous pressure aquifer, ________ of the total thickness of water bearing formation is screened.

    a. Central 70 – 80%
    b. Top 50%
    c. Bottom 50%
    d. 90%

59. Maximum possible drawdown for a well installed in a confined aquifer is the distance between original piezometric level and ________.

    a. Top of the aquifer at well face
    b. One-meter above the top of the aquifer
    c. Centre of the confined aquifer
    d. Pumping water level

60. In order to install a well in unconfined aquifer, preferably ______ of the aquifer is screened.

    a. Bottom one-third
    b. Bottom one-half
    c. Bottom one-third to one-half
    d. Central one-third

61. The overlapping of cone of depressions of two wells influences the _______.

    a. Discharge rates of both the wells
    b. Drawdowns of both the wells
    c. Both (a) and (b)
    d. Discharge rate of well with lesser yield

62. In alluvial plains, the grouping of wells should not be made at intervals less than _______.

    a. 70 m
    b. 100 m
    c. 50 m
    d. 150 m

63. For keeping the head losses to a minimum, water velocity through the individual openings of the well screen should range between _______.

    a. 3.0 to 7.5 cm/s
    b. 1.5 to 5.0 cm/s
    c. 1.0 to 3.5 cm/s
    d. 0.5 to 6.0 cm/s

64. The slot width of the screen is decided on the ________.

    a. Basis of grain size distribution of formation
    b. Basis of structural strength of screen
    c. Basis of entrance velocity through screen
    d. Both (b) and (c)

65. Slot size of screen should range between ________ to avoid its clogging.

    a. 1.5 to 5 mm
    b. 3.0 to 6.0 mm
    c. 1.0 to 3.0 mm
    d. 3.0 to 4.5 mm

66. Even after the proper well development, on an average ______ of the open area of well screen remains clogged by the aquifer materials.

    a. 75%
    b. 50%
    c. 1/3rd
    d. 1/4th

67. In India, the slotted pipes are manufactured with the slot sizes of widths _____.

    a. 1.6 mm and 3.2 mm
    b. 1.8 mm and 3.6 mm
    c. 3.2 mm and 4.5 mm
    d. On demand

68. In India, slots of a well screen are manufactured with a minimum spacing of _______.

    a. 3 mm
    b. 4.5 mm
    c. 1.2 mm
    d. 3.2 mm

69. The slot openings' width of a well screen is decided on the basis that about _____ of formation material can pass through it.

    a. 70%
    b. 50%
    c. 90%
    d. 30%

70. In Ashim filter, _________.

    a. Perforations of size 1 cm are recommended
    b. Continuous slots are made
    c. Gravel pack is casted on a slotted pipe
    d. Both (a) and (c)

71. The formation of a thin layer of coarse material around the well screen during the well development is called a _____.

    a. Gravel packing
    b. Natural gravel packing
    c. Gravel shrouding
    d. None of the above

72. The removal of fine sand and silt from the aquifer through well screen by surging and bailing action is known as__________.

a. Well development
b. Natural gravel packing
c. Formation cleaning
d. All above

73. ________ refers to the particle size where 10% of sand is finer.

a. D_{90}
b. D_{10}
c. $D_{10\text{-}90}$
d. None of the above

74. P-A ratio is the ratio of _________ sieve sizes in sieve analysis of granular material.

a. D_{60} to D_{10}
b. D_{40} to D_{90}
c. D_{10} to D_{60}
d. D_{30} to D_{70}

75. For designing gravel pack, the term uniformity coefficient for granular materials was proposed by ______.

a. Hazen
b. William
c. Dupuit
d. Forchheimer

76. The Hazen's uniformity coefficient for fairly even grained sand will be between ________.

a. 2-3
b. 1-2
c. 1-3
d. None of the above

77. Wells installed in the formations having effective size of ______ does not require gravel pack.

a. 0.25 mm
b. 0.17 mm
c. 0.09 mm
d. 0.1 mm

78. Gravel pack around the well screen _______.

a. Increases incrustation
b. Decreases incrustation
c. Increases effective radius of well
d. Both (b) and (c)

79. A good gravel pack material should be insoluble siliceous material with less than _______% limestone.

a. 1
b. 10
c. 5
d. 2.5

80. As per the recommendations of American Society of Agricultural and Biological Engineers, the maximum size of the particle in a gravel pack should not exceed ________.

a. 3.2 mm b. 6.4 mm
c. 5.0 mm d. 2.1 mm

81. The pack-aquifer ratio is usually the ratio of _______ size of gravel pack and formation material particles.

a. 60% b. 40%
c. 50% d. 70%

82. For keeping the minimum head loss through the gravel pack, the pack-aquifer ratio should be _______.

a. 9.0 b. 4.0
c. 12.0 d. 6.0

83. For stable filtering action, the upper limit of pack-aquifer ratio in gravel pack should be ________.

a. 6.0 b. 9.0
c. 12.0 d. 4.0

84. In practical situations, gravel pack thickness should not be less than ________.

a. 20 cm b. 15 cm
c. 0.3 m d. 5 cm

85. Generally, the upper limit of gravel pack thickness is kept around _______.

a. 10 cm b. 0.3 m
c. 15 cm d. 0.5 m

86. The name of the scientist to study the flow in a multiple well points system first was ______.

a. Huisman b. Darcy
c. Forchheimer d. Kirkham

87. To augment yield in an open well, the boring of lateral tunnels in semi-consolidated formations below water table are called ________.

a. Lateral holes b. Revitalization holes
c. Horizontal gallery d. None of the above

88. The depth of driven and jetted tube wells is generally not over ________.

    a. 30 m
    b. 15 m
    c. 50 m
    d. 100 m

89. The size of driven tube wells usually ranges between _______.

    a. 1.0-3.0 inch
    b. 2.0-5.0 inch
    c. 3.0-6.0 inch
    d. ¾ - 2.0 inch

90. The construction of tube wells by driving into unconsolidated formations is generally used for _______.

    a. Domestic water supply
    b. Irrigation
    c. Both (a) and (b)
    d. None of the above

91. Percussion method of drilling is also known as _______.

    a. Rotary method
    b. Cable tool method
    c. Reverse circulation rotary method
    d. Air percussion method

92. The part of percussion rig used for raising and dropping of bailer is called a ______.

    a. String
    b. Mast
    c. Derrick
    d. Spudder

93. The process of recovering lost or jammed up drilling tools in borehole is known as ________.

    a. Trapping
    b. Fishing
    c. Recovering
    d. Both (a) and (c)

94. A total weight of tools that a light duty percussion rig can handle, ranges between ______.

    a. 200-700 Kg
    b. 500-1500 Kg
    c. 150-500 Kg
    d. 350-800 Kg

95. Drilling of holes up to a depth of _______ can be drilled suitably by light duty percussion rigs.

    a. 130 m
    b. 200 m
    c. 300 m
    d. 50 m

96. Drilling of holes up to a depth of ______ can be drilled by medium range percussion rigs.

a. 300 m
b. 200 m
c. 350 m
d. Any dimensions

97. The tool-weights handling capacity of medium range percussion rigs range between ______.

a. 350-800 Kg
b. 500-1000 Kg
c. 1000-1700 Kg
d. 800-1500 Kg

98. In direct rotary drills, for pumping drill mud at pressures more than about ______ Kg/cm^2, the double acting piston type pump is used.

a. 9.0
b. 3.5
c. 4.0
d. 6.0

99. ______ bits are used for drilling in soft and unconsolidated formations using direct rotary drilling.

a. Rolling cutter
b. Button
c. Flat-fluted
d. Drag

100. Fish tail bit is a ____________ ,which is used in a rotary drill.

a. Three-way drag bit
b. Three-way roller bit
c. Two-way drag bit
d. Two-way roller bit

101. A __________ in a rotary drill is also called a pilot bit.

a. Fish-tail bit
b. Three-way bit
c. Two-way bit
d. Six-way bit

102. A drill stem is a __________, which operates the drill bit in rotary drilling method.

a. Solid rotating shaft
b. Hollow rotating shaft
c. Pipe for conveying drill cuttings
d. None of the above

103. In direct rotary drilling, the drill speed usually ranges between ______ rpm.

a. 30-150
b. 50-200
c. 100-300
d. 60-100

104. The drilling fluid velocity in annular space for direct rotary drilling method usually ranges between ________ m per min.

a. 40-60
b. 60-80
c. 20-30
d. 80-100

105. The rubbery lining of silt clay and colloids formed on the wall of the borehole during direct rotary drilling is called ____________.

a. Lining cake
b. Filter cake
c. Mud cake
d. Both (b) and (c)

106. A borehole will remain open as long as the hydrostatic pressure of drilling fluid exceeds ________.

a. Earth pressure
b. Artesian pressure
c. Bit pressure
d. Both (a) and (b)

107. For quickly sealing the sides of borehole, high-grade ________ drilling mud is used.

a. Bentonite
b. Kaolinite
c. Montmorillonite
d. Illite

108. Among the following, _______ is not the fishing tool in direct rotary drilling.

a. Tapered tap
b. Die overshot
c. Circulating slip overshot
d. Fishing hook

109. For drilling tube wells with a discharge capacity of about ________, light direct rotary rigs are used.

a. 30 l/s
b. 50 l/s
c. 120 l/s
d. 80 l/s

110. For drilling tube wells with a discharge capacity range of ________, medium size direct rotary rigs are used.

a. 30-60 l/s
b. 120-150 l/s
c. Less than 40 l/s
d. More than 150 l/s

111. For drilling tube wells with a discharge capacity range of ________, heavy-duty direct rotary rigs are used.

    a. 60-150 l/s
    b. 150-200 l/s
    c. 120-150 l/s
    d. 60-90 l/s

112. The tube wells up to depths of __________ can be drilled in alluvial and semi-consolidated formations by direct rotary drilling.

    a. 200 m and above
    b. 225 m and above
    c. 300 m and above
    d. None of the above

113. ________ is used to circulate drilling fluid for drilling holes deeper than 300 m or more using reverse circulation rotary drilling method.

    a. Compressed air
    b. High capacity centrifugal pump
    c. Piston pump
    d. Gravity head

114. The size of the drill pipe is usually ________ in direct rotary drill as compared to reverse rotary drill.

    a. Larger
    b. Same
    c. Smaller
    d. None of the above

115. ________ pump is used for pumping the drilling fluid in reverse circulation rotary drill.

    a. Piston
    b. Centrifugal
    c. Rotary
    d. Gear

116. For drilling holes of size _______ cm and above, reverse circulation rotary drills are preferred.

    a. 10
    b. 25
    c. 35
    d. 15

117. The water requirement in reverse circulation rotary drill is ______ times the requirement in direct rotary drill.

    a. 5
    b. 2
    c. 0.5
    d. 1.0

118. The construction of gravel pack tube wells is suitable with _______ rotary drills.

a. Air percussion
b. Direct
c. Air
d. Reverse circulation

119. While drilling with air, its velocity in the annular space must be around _______ m/min to clear the drill cuttings from borehole.

a. 900
b. 500
c. 1500
d. 250

120. A tool name *down the hole hammer* is used in _______ drilling of well.

a. Direct rotary
b. Air percussion rotary
c. Reverse circulation rotary
d. Percussion

121. In air percussion rotary drilling, the hammer hits the bit while it is being rotated at a rate of _______ blows per min.

a. 500-700
b. 600-1200
c. 1000-2000
d. 150-300

122. The best method for obtaining uncontaminated samples of underground formations is______.

a. Core drilling
b. Rotary drilling
c. Percussion method
d. Auger method

123. The holes for observation wells and piezometers are drilled using _______ drilling.

a. Percussion
b. Rotary
c. Core
d. Water jet

124. Drilling of crooked holes can be avoided by keeping the diameter of the borehole less than _______ times the diameter of the drill pipe.

a. 1.5
b. 2
c. 2.5
d. 1.75

125. By maintaining speed of bit rotation at least ________ rpm, drilling of crooked holes can be avoided.

a. 200
b. 250
c. 175
d. 150

126. The problem of bailing up can be avoided by ________.

a. Rotating the bit at high speed
b. Adapting high degree of penetration
c. Frequent spudding of drill pipe
d. Washing the bit

127. The problem of ________ does not occur if enough time is given to the circulating fluid to mix with drill cuttings.

a. Bailing up
b. Lost circulation
c. Crooked hole development
d. None of the above

128. When a water bearing formation with higher hydrostatic pressure than drilling mud column is encountered during drilling process, a substantial ________ in returning circulation fluid occurs.

a. Loss
b. Decrease
c. Increase
d. None of above

129. The cutting shoe in case of cavity bores should be left about ________ above the bottom of clay layer.

a. 100 cm
b. 60 cm
c. At least 30 cm
d. 50 cm

130. Before lowering the well assembly in a borehole, its bottom end is closed with a cap named as ________.

a. Bail plug
b. End plug
c. Bottom plug
d. Pipe plug

131. Drill stem test is usually carried out to ________________.

a. To know the groundwater quality
b. Demarcate saline aquifer formation
c. To know permeability of aquifer
d. Both (a) and (b)

132. Well development using surging method normally completes in ________.

a. Less than 1 day
b. 2-3 days
c. More than 8 hours
d. 16-20 hours

133. Rawhiding the well is a development technique by _______.

a. Pumping
b. Surging
c. Back-washing
d. None of the above

134. A well development procedure in which pump is started and stopped intermittently is known as ________.

a. Surging
b. Start-stop method
c. Rawhiding
d. None of above

135. Well development under severe sand pumping conditions may cause ________ of the pump.

a. Sand locking
b. Abrasion
c. Both (a) and (b)
d. None of the above

136. The wells, where depth of water exceeds ________ of the total depth of the well, surging method for well development by using compressed air can be successfully used.

a. 2/3
b. 1/3
c. ½
d. ¼

137. The tendency of the drilling mud to stick to formation sand grains can be counteracted with the use of ________.

a. Sodium chloride
b. Monophosphates
c. Gypsum
d. Polyphosphates

138. To accomplish well development by dispersing agents, about _____ kg of polyphosphate is required to add in 100 l of water.

a. 1.5
b. 1.0
c. 0.5
d. 0.25

139. The yield is almost ________ to the drawdown for the wells tapping greater thickness of water bearing formations.

a. In inverse proportion
b. In direct proportion
c. Square root
d. None of the above

140. __________ is a plot between discharge vs drawdown for a well.

a. Performance curve
b. Type curve
c. Well characteristics curve
d. Capacity-drawdown curve

141. The build-up of the material around the well screen or the metallic well section to form an impervious wall is known as ________.

a. Incrustation
b. Slime
c. Junk
d. Screen blocking

142. Slight fluffy carbonate deposit on well screen is termed as _________, which can be removed easily.

a. Hard incrustation
b. Soft incrustation
c. Slime formation
d. Rusting

143. Water-soluble sulphates cause ________ on well screen, which requires continuous treatment.

a. Hard incrustation
b. Soft incrustation
c. Rusting
d. Slime formation

144. In acid treatment of incrustation problem, ________ containing a suitable inhibitor is used.

a. Hydrochloric acid
b. Sulphuric acid
c. Polyphosphate
d. Chlorine

145. To keep iron in solution during acid treatment for incrustation problem, ________ is a stabilizer added to hydrochloric acid.

a. Calcium carbonate
b. Chlorine
c. Sodium chloride
d. Rochelle salt

146. A strong oxidizing agent, which kills bacteria and organic slime responsible for clogging of wells, is ________.

a. HCl
b. Polyphosphate
c. Chlorine
d. H_2SO_4

147. For breaking up the incrustations of well screens, ________ are used.

a. Polycarbonates
b. $CaCO_3$
c. Polysulphates
d. Polyphosphates

148. Dissolved salts in groundwater in excess of ________ mg/l may cause serious corrosion.

a. 1000 b. 500
c. 5000 d. 4000

149. Carbon dioxide accelerates the corrosion process if it is in excess of ________ mg/l.

a. 65 b. 50
c. 40 d. 25

150. Under acidic conditions, chlorine in excess of ________ mg/l is a corrosion accelerator for most of the metals.

a. 400 b. 300
c. 350 d. 275

151. A geological formation having enough capacity to store and transmit water is called ________.

a. Aquiclude b. Aquitard
c. Aquifuse d. Aquifer

152. The formations having high porosity but low capability to transmit water are known as ________.

a. Aquiclude b. Aquitard
c. Aquifer d. Aquifuse

153. The geological formations, which can neither store nor transmit water are called ________.

a. Aquitard b. Aquiclude
c. Aquifuse d. Clay formation

154. The geological formation with too small hydraulic conductivity to permit well development is known as ________.

a. Aquifuse b. Aquiclude
c. Aquitard d. Both (b) and (c)

155. A false aquifer is also known as ________.

a. Wrong positioned aquifer b. Perched aquifer
c. Very large aquifer d. Both (a) and (c)

156. The locations where the aquifer is overlain by a stiff clay layer of _______ thickness are suitable for the development of cavity wells.

a. 3-5 m
b. 1-2 m
c. 0.5-1.0 m
d. None of the above

157. For the development of stabilized cavity well, around ______ of sand comes out of borehole.

a. 6-9 m^3
b. 12-18 m^3
c. 3-6 m^3
d. 6-12 m^3

158. __________ are installed using pull back method.

a. Well casings
b. Well screens
c. Well pipes
d. Gravel packing

159. The frictional head loss due to movement of water within an aquifer is called ________.

a. Formation loss
b. Well loss
c. Screen loss
d. Aquifer loss

160. The head loss caused due to the flowing of water through well pipe is known as ________.

a. Formation loss
b. Aquifer loss
c. Well loss
d. Screen loss

161. An aquifer bounded from above and below by aquicludes or aquifuses is known as ________.

a. Perched
b. Unconfined
c. Semi-confined
d. Confined

162. If properties of a medium do not change with respect to direction, it is termed as ________.

a. Heterogeneous
b. Anisotropic
c. Isotropic
d. Homogeneous

163. If properties of a medium vary depending upon the direction in which these are measured, it is termed as ________.

a. Isotropic
b. Anisotropic
c. Homogeneous
d. Heterogeneous

164. No change in the properties of isotropy or anisotropy throughout the medium is termed as ________.

a. Anisotropic
b. Heterogeneous
c. Isotropic
d. Homogeneous

165. The variation in the properties of isotropy or anisotropy throughout the medium is termed as ________.

a. Heterogeneous
b. Homogeneous
c. Isotropic
d. Anisotropic

166. In a ________ medium, the permeability of a medium is same in any direction from a point and is constant throughout the medium.

a. Heterogeneous anisotropic
b. Homogeneous anisotropic
c. Homogeneous isotropic
d. Heterogeneous isotropic

167. In a ________ medium, the permeability of a medium varies depending upon the direction from a point and variation is constant throughout the medium.

a. Heterogeneous anisotropic
b. Homogeneous anisotropic
c. Homogeneous isotropic
d. Heterogeneous isotropic

168. In a ________ medium, the permeability of a medium is same in any direction from a point, but varies spatially throughout the medium.

a. Heterogeneous anisotropic
b. Homogeneous anisotropic
c. Homogeneous isotropic
d. Heterogeneous isotropic

169. In a ________ medium, the permeability of a medium varies depending upon direction from a point and variation is not constant throughout the medium.

a. Heterogeneous anisotropic
b. Homogeneous anisotropic
c. Homogeneous isotropic
d. Heterogeneous isotropic

170. If the long-term groundwater withdrawal from the aquifer exceeds its recharge, the aquifer is said to be in ________ condition.

a. Under-draft
b. Overdraft
c. Negative draft
d. None of the above

171. The Darcian velocity of groundwater flow under the unit hydraulic gradient will be equal to ________.

a. Transmissivity
b. Specific discharge
c. Hydraulic conductivity
d. None of above

172. In actual field conditions, the water table slopes typically are small and are in the range of ________.

a. 0.5 to 1.0%
b. 0.1 to 1.0%
c. 1.0 to 2.0%
d. None of above

173. Even after well development, ________% of the total open area of the screen remains blocked by aquifer material or gravel pack.

a. 35
b. 25
c. 60
d. 50

174. Out of the total annual rainfall of 400 million ha-m in India, around ________ million ha-m is stored as soil moisture.

a. 113
b. 150
c. 165
d. 75

175. Approximately, ________% of the total annual precipitation in India percolates down to reach the groundwater table.

a. 7.5
b. 12.5
c. 25.5
d. 3.0

176. Annually, around ________ million ha-m of water from all rivers in India seep to join groundwater.

a. 5
b. 9
c. 7
d. 16

177. About ________ million ha-m of water is added to groundwater annually in India by irrigated fields through deep percolation.

a. 23
b. 57
c. 45
d. 50

178. Excluding the soil moisture storage from annual precipitation, ________ million ha-m meets as the total groundwater recharge in India.

a. 412
b. 57
c. 180
d. None of the above

179. An art of locating underground water by involuntary muscular reaction on the part of an individual is known as ________.

a. Muscular exploration
b. Water sensitivity
c. Water divining
d. None of the above

180. To be used as casing pipe, the tensile strength of PVC pipes should be greater than ________ kg/cm^2.

a. 325
b. 175
c. 250
d. 550

181. The ratio between diameter of PVC pipe and its ________ is known as its standard dimension ratio (SDR).

a. Length
b. Thickness
c. Diameter of borehole
d. None of above

182. The standard dimension ratio for the installation of PVC pipe as a casing and screen for tube wells should range between ________.

a. 10-20
b. 15-30
c. 16-28
d. 5.0-7.5

183. The strength of tube well screens of PVC material with horizontal slots is ________ that of the casings.

a. 0.5-0.7 times
b. 0.75-1.00 times
c. 1.0-2.0 times
d. Same

184. An aquifer with a dissolved solids content of 600 mg/l is showing certain electrical resistivity. If we consider that dissolved solids content has decreased to 300 mg/l, then the electrical resistivity will ________.

a. Become 2-times of the original
b. Become half of the original
c. Become 4-times of the original
d. Not change

185. Consider that the porosity of 1 m^3 saturated sand is 35%. If 0.17 m^3 of water is drained from this saturated sand under gravitational conditions. Its specific retention will be

a. 18%
b. 0.18 m^3
c. 52%
d. Either (a) or (b)

186. A water table well of 10 cm diameter is installed in a 25 m thick aquifer and is pumping water at a uniform rate of 1 m^3/min. Drawdown values of 6 m and 0.7 m are monitored in the observation wells at distances of 1 m and 50 m from well, respectively. The hydraulic conductivity of the aquifer is

a. 0.001 m/min.
b. 0.0054 m/min.
c. 0.01 m/min.
d. 0.054 m/min.

187. An 8 cm diameter well is installed in a 10 m thick water-bearing formation under confining conditions and is pumping the aquifer with a constant rate of 80 l/min. Two observation wells are installed at distances of 7 and 30 m from the well centre in which the values of the drawdowns monitored are 3.2 and 0.1 m, respectively. The transmissivity of the aquifer is

a. 0.00798 m^2/min.
b. 0.0598 m^2/min.
c. 0.00598 m^2/min.
d. 0.01 m^2/min.

188. In a Darcy apparatus, a column of diameter 0.10 m is filled with clean coarse sand of porosity 35%. A uniform flow rate of 0.5 l/s was established through this column. The average interstitial velocity through column will be

a. 0.182 m/s
b. 1.82 m/s
c. 0.0182 m/s
d. 0.20 m/s

189. A tube well has a designed yield of 150 m^3/h. Aquifer to be tapped is 25 m thick. It was decided to use a screen with length of 20 m. Considering the 20% open area of the screen and 50% effective open area, diameter of the screen will be

a. 220 mm
b. 250 mm
c. 200 mm
d. 150 mm

190. Which of the following is not a fishing tool?

a. Tapered tap
b. Die overshot
c. Fishing overshot
d. Circulating slip overshot

191. The blade angle of a propeller depends upon the

a. Working head
b. Speed of impeller
c. Both head and speed
d. None of the above

192. Because of the steep horsepower curve at shut-off, propeller and mixed flow pumps are started ________.

a. Against an open discharge
b. Against a pressurized system
c. At closed discharge valve
d. None of the above

193. Cavitation phenomenon occurs in the ________

a. Centrifugal pumps
b. Turbine pumps
c. Propeller pumps
d. All above

194. The efficiency of the jet pump is influenced by the

a. Nozzle throat ratio
b. Capacity of the centrifugal pump
c. Depth of the tapped aquifer
d. Nozzle size

195. The spacing between nozzle and throat of the jet pump is equal to

a. Two nozzle diameter
b. One nozzle diameter
c. Nozzle circumference
d. None of above

196. Eductor pipe is a component related to

a. Jet pump
b. Propeller pump
c. Mixed-flow pump
d. Airlift pump

197. In airlift pump, the eductor and air pipes must be submerged in water in the well with ______ or more of their lengths extending below the pumping level.

a. 50%
b. 30%
c. 10%
d. 80%

Objectives based upon the chapters 5 to 11

Fill in the blanks

1. Crop production in poor drained soils decreases because of_______.

   a. Limited amount of soil b. Lack of nutrients
   c. Non-availability of nitrogen d. All above

2. The diffusion of gases _______ in wet soils.

   a. Is extremely fast b. Is extremely slow
   c. Ceases d. Beyond control

3. Under waterlogged conditions, the anaerobic decomposition of organic matter results in the production of _______ gas.

   a. Methane b. H_2O_2
   c. Cl_2 d. NH_3

4. In waterlogged soils, mineral substances are altered _______.

   a. From oxidized state to reduced state b. From reduced state to oxidized state
   c. From carbonates to bicarbonates d. From sulphates to bisulphates

5. The rate of decomposition of organic matter ________ in waterlogged soils.

   a. Increases b. Decreases
   c. Remains same d. None of these

6. ________ conditions restrict the ready warm up of soils.

   a. High water table b. Low water table
   c. Higher moisture conditions d. Both (a) and (c)

7. The depth to water table from ground surface can be observed with the help of a /an_______.

   a. Observation well
   b. Piezometer
   c. Tensiometer
   d. All above

8. For checking artesian pressure, piezometers are _______.

   a. Installed at same depths at nearby locations
   b. Installed at variable depths at nearby locations
   c. Installed in the adjoining aquifers
   d. None of these

9. Artesian pressure is indicated in the aquifer, if _______.

   a. Water level in the deeper piezometer rises to greater height than shallower piezometer
   b. Water level in the deeper piezometer is lower than the level in piezometer at lesser depth
   c. Water stands at the same level in the piezometers
   d. None of these

10. From drainage point of view, if the permeability of subsoil is about _______ that of the surface soil it can be considered as impermeable.

    a. 1/10th
    b. 1/5th
    c. Half
    d. 1/3rd

11. The volume of water, which can be drained per unit volume of soil when soil moisture pressure is decreased from atmospheric to some specified negative pressure is known as _______.

    a. Gravitational volume
    b. Specific yield
    c. Drainable pore volume
    d. None of these

12. The drainable porosity at saturation is _______.

    a. Zero
    b. 100%
    c. 50%
    d. Not measurable

13. The flow of fluids through __________ is governed by Poiseuille's law.

    a. Macro pores
    b. Well bores
    c. Capillary tubes
    d. None of these

14. The basis of a number of equations used for determining hydraulic conductivity from pore size distribution of soil is _______.

a. Ohm's law
b. Darcy's law
c. Newton's law
d. Poiseuille's law

15. The fluid flow through the capillary varies as the ______ of its radius.

a. Square
b. Cube
c. Square root
d. Cube root

16. Darcy's law for flow of water through soils is synonymous with _______.

a. Ohm's law
b. Fick's law
c. Fourier law
d. All above

17. The law developed by Henry Darcy was proposed in the year _______.

a. 1856
b. 1890
c. 1850
d. 1857

18. As per _______ law, the flow of water through porous material is proportional to the hydraulic gradient.

a. Ficks'law
b. Ohm's law
c. Poiseville's law
d. Darcy's law

19. The energy, a mass possesses due to its position above some arbitrary reference plane is equal to _______.

a. Gravitational energy
b. Potential energy
c. Kinetic energy
d. Both (a) and (b)

20. For the applicability of Darcy's law, the flow of water through the porous medium must be in the ________ regime.

a. Laminar
b. Non-laminar
c. Steady
d. Unsteady

21. Intrinsic permeability has the dimensions of _______.

a. L^2
b. L
c. L^3
d. None of these

22. With the increase in capillary pressure, the capillary conductivity _______.

    a. Remains same
    b. Increases
    c. Decreases
    d. Tends to sharp increase

23. The capillary conductivity at a capillary pressure of about _____ is approximately equal to zero.

    a. 1/10 to 1/3 atm
    b. At 1 atm
    c. 1/3 to 1 atm
    d. None of the above

24. The thickness of capillary fringe will be _________ in clayey as compared to sandy soils.

    a. Variable
    b. Less
    c. Equivalent
    d. More

25. A plot of equi-potentials and of streamlines is known as _______.

    a. Flow net
    b. Equilines
    c. Flow curve
    d. Flow path

26. The flow occurs _______ to the equipotential lines in an isotropic media.

    a. Parallel
    b. At an angle of 90°
    c. Depending upon the position from drain
    d. Independent

27. The lines of flow are called _______ to equipotential lines.

    a. Streamlines and are orthogonal
    b. Streamlines and are parallel
    c. Streamlines and are independent
    d. None of these

28. The auger hole method measures the ________ of the hydraulic conductivity of soil.

    a. Three-dimensional
    b. Vertical component
    c. Horizontal component
    d. None of these

29. _______ component of hydraulic conductivity is important from drainage point of view.

a. Three-dimensional
b. Vertical
c. Horizontal
d. Any of above

30. The single auger hole method for determining soil hydraulic conductivity is developed by _______.

a. Kirkham
b. Ernst
c. Luthin
d. Hooghoudt

31. Under actual field conditions, hydraulic conductivity can be as low as _______ m/d.

a. 0.01
b. 0.001
c. 0.1
d. 0.009

32. Under actual field conditions, the hydraulic conductivity in the field can exceed even _______ m/d.

a. 100
b. 10
c. 1
d. 5

33. _______ studied flow into an auger hole and developed charts.

a. Ernst
b. Hooghoudt
c. Kirkham
d. Richards

34. For determining hydraulic conductivity of two layered soil using auger hole method, the bottom of the first hole should be approximately _______ above the lower layer.

a. 100 cm
b. 10 cm
c. 0.1 m
d. Two times the diameter of second hole

35. To determine the hydraulic conductivity of layered soils using auger hole method, _______ equation can be used.

a. Hooghoudt's
b. Childs
c. Ernst's
d. Richards

36. The method for determining hydraulic conductivity using the two auger holes is proposed by _______.

    a. Ernst's | b. Childs
    c. Hooghoudt's | d. Kirkham

37. To determine hydraulic conductivity in the presence of water table, the _______ method is used.

    a. Single-auger-hole | b. Two-auger-hole
    c. Pipe-cavity | d. All above

38. To determine hydraulic conductivity in the absence of water table, the _______ is used.

    a. Shallow well pump-in test method | b. Permeameter method
    c. Pond-infiltration test method | d. All above

39. To reduce lateral and seepage loss per unit area of infiltration pond, the infiltration test is conducted in a _______ pond.

    a. Circular | b. Rectangular
    c. Square | d. None of these

40. A hydraulic gradient of value ______ is required for gravity flow.

    a. 980 cm/cm | b. Unity
    c. 9.81 m/cm | d. Both (a) and (c)

41. The lines joining the points of equal elevation of groundwater level are called _______.

    a. Isopaths | b. Isohyets
    c. Isobaths | d. None of these

42. If two piezometers tapping different depths in the aquifer give _____, there is no vertical flow component.

    a. Difference more than 10 cm | b. Same levels
    c. Difference less than 10 cm | d. None of these

43. The higher water level in the deeper piezometer than the level in the piezometer at shallower depth at the same point indicates _______ direction of flow.

   a. Upward b. Downward
   c. Horizontal d. No flow

44. The depth of water to be removed from an area in a day is known as _______.

   a. Drainage coefficient b. Drainage depth
   c. Capacity of drainage system d. All above

45. The pipe of a _______ is not perforated.

   a. Monitoring well b. Piezometer
   c. Observation well d. All above

46. Piezometers are used to measure the _______ in the penetrated aquifer.

   a. Water level b. Water table
   c. Hydraulic head d. Both (a) and (c)

47. The rate of rise of water in an auger hole is _______ to the circumference of the hole.

   a. Two-times b. Inversely proportional
   c. Four-times d. Directly proportional

48. The rate of water rise in the auger hole is _______ proportional to the cross sectional area of the hole.

   a. Inversely proportional b. Directly proportional
   c. Half-times d. Three-times

49. For Darcy's law to be valid, Reynold's number should be _______.

   a. Less than unity b. Equal to unity
   c. Equal to 10 d. Between 1 to 10

50. _______ removes gravitational water from below the root zone and creates aeration.

   a. Evaporation b. Surface drainage
   c. Subsurface drainage d. Both (b) and (c)

51. If the porosity of different soil samples is same, then a soil with _______ has higher hydraulic conductivity.

    a. Greater number of macro pores
    b. Greater number of micro pores
    c. Less number of macro pores
    d. Less number of micro pores

52. Soils having high _______ content undergo deterioration of soil structure with application of irrigation water and is responsible for permeability decrease.

    a. NH_4
    b. Na
    c. Ca
    d. K

53. The pH of sodic soils is _______.

    a. < 7.0
    b. Ranging between 7.0 to 8.5
    c. > 8.5
    d. Not within defined range

54. For saline soils, the electrical conductivity of saturation extract is _______.

    a. > 4 dS/m
    b. 15 dS/m
    c. < 4 dS/m
    d. Between 4 and 6 dS/m

55. The exchangeable sodium percentage of saturation extract of saline soil is _______.

    a. < 7.5%
    b. < 15 dS/m
    c. < 15%
    d. < 4%

56. The pH of saturation extract of saline soil is generally _______.

    a. Less than 7.0
    b. > 8.5
    c. Lies between 6.0 to 8.5
    d. < 8.5

57. Saline sodic soils have _______.

    a. EC greater than 4 dS/m
    b. EC > 4 dS/m and ESP < 15%
    c. ESP greater than 15%
    d. Both (a) and (c)

58. A drainage coefficient of approximately _______ mm/d is appropriate in humid areas.

    a. 10
    b. 3
    c. 100
    d. 1

59. A drainage coefficient of about _______ mm/d is appropriate in arid areas.

a. 100 b. 3
c. 10 d. 1

60. _______ drains must be placed in or above an area of local water concentration to remove excess water appearing at that point.

a. Seepage b. Relief
c. Interceptor d. None of these

61. To determine the interconnected void space in the soil, its _______ is used.

a. Saturation percentage b. Porosity
c. Specific yield d. Drainable porosity

62. In surface drainage systems, _______ shape open drains are usually constructed.

a. Rectangular b. Trapezoidal
c. Triangular d. Square

63. The ploughing of land to form a series of low narrow ridges separated by parallel dead furrows with ridges oriented along greatest land slope is known as _________ system of drainage.

a. Diversion-ditch b. Random
c. Bedding d. Interception

64. _______ is a bedding technique in which a series of broad, low ridges parabolic convex shaped beds are developed with the ridges midway between drains.

a. Crowning b. Furrowing
c. Wide-crowning d. Narrow-furrowing

65. Seepage from canals can be collected conveniently by laying _______ drains.

a. Interceptor b. Relief
c. Seepage d. Surface

66. To check lateral flow of groundwater from irrigated areas at higher elevations, _______ drains can be used.

a. Relief b. Seepage
c. Surface d. Interceptor

67. _______________ are known as relief mole drains system.

a. Combination of deep lateral and shallow mole drains
b. Combination of shallow lateral and shallow mole drains
c. Combination of deep lateral and deep mole drains
d. Combination of shallow lateral and deep mole drains

68. In most practical situations, mole drains are constructed at a depth of _______.

a. < 0.5 m
b. 0.7 m
c. 1.5 m
d. Greater than 2 m

69. Generally, the efficient service life of mole drains is _______ years.

a. Greater than 10
b. Up to 5
c. Up to 10
d. Less than 5

70. As per USDA-SCS, minimum grade criteria for the areas having no sedimentation hazard, the velocity of flow in subsurface drains should not fall below _______.

a. 0.1 m/s
b. 0.15 m/s
c. 0.01 m/s
d. 1.5 m/s

71. As per USDA-SCS, minimum grade criteria for the areas having sedimentation hazard, the velocity of flow in subsurface drains should not fall below _______.

a. 0.45 m/s
b. 0.15 m/s
c. 0.01 m/s
d. 2.0 m/s

72. Drainage system will be most effective when laterals are installed _______ to water table contours.

a. Across
b. Perpendicular
c. Parallel
d. Diagonal

73. _______ are provided at the locations either where two mains meet or where two or more laterals connect to a collector line.

a. Junction box on manholes
b. Air vent
c. Sedimentation basins
d. Both (a) and (c)

74. To avoid the danger from impact loads imposed by heavy vehicles, drain depth should not be less than _______.

    a. 1 m   b. 0.5 m
    c. 1.5 m   d. 2.0 m

75. _______ pumps are most commonly used in surface drainage systems.

    a. Jet   b. Propeller
    c. Piston   d. Centrifugal

76. The _______ pump do not require priming.

    a. Turbine   b. Centrifugal
    c. Jet   d. Axial flow

77. The pumps with the combined features of axial flow and radial flow pumps are called _______.

    a. Mixed flow pumps   b. Turbine pumps
    c. Propeller pumps   d. Volute pumps

78. In case, the drainage line outlet is below the water level in main drain ________is used.

    a. Pump outlet   b. Flap gates
    c. Evaporation sump   d. All above

79. The aquifer tapped by the drainage well should ________.

    a. Have sufficient area   b. Be covering the area to be drained
    c. Have Characteristics which promote large pumping   d. All above

80. Return flow can consist of ___________.

    a. Surface runoff from irrigation   b. Subsurface drainage effluent from deep seepage
    c. Both (a) and (b)   d. None of these

81. Leaching fraction is that part of applied water, which _________.

    a. Percolates through the vadose zone   b. Percolates through the crop root zone
    c. Percolates to the deeper formations   d. All above

82. Generally, the subsurface return flows in humid areas are ________ in nature.

 a. Saline
 b. Marginal
 c. Good
 d. Brackish

83. The production of irrigation return flow can be reduced by ________.

 a. Increasing irrigation application efficiency
 b. Increasing water conveyance efficiency
 c. Increasing water efficiency
 d. All above

84. The production of irrigation return flow can be reduced by ________.

 a. Increasing surface drainage
 b. Reducing leaching
 c. Reducing water application rate
 d. All above

85. In comparison to applied irrigation water, the salinity level of subsurface return flow _______.

 a. Increases
 b. Decreases
 c. Remains same
 d. Varies continuously

86. As, _______ is very soluble in water, it can be expected in both surface and subsurface drainage waters.

 a. Nitrogen
 b. Oxygen
 c. Chlorine
 d. Hydrogen

87. While deriving the Hooghoudt's analysis, it is assumed that _______ at any point is equal to the slope of water table above that point.

 a. Hydraulic head
 b. Hydraulic gradient
 c. Flow rate
 d. Water declining rate

88. According to _______ assumption, the hydraulic gradient at any point is equal to slope of water table above that point.

 a. Dupuit-Forchheimer
 b. Ernst's
 c. Richard's
 d. Kirkham's

89. In the ellipse equation, as the vertical distance from the bottom of drain to impermeable layer approaches infinity, the spacing between drains approaches _______.

a. Zero
b. Infinity
c. Equal to depth of drain from the surface
d. None of these

90. The Hooghoudt's equation reduces to _______ equation, if drain is considered empty.

a. Parabolic
b. Circle
c. Ellipse
d. None of these

91. The semi-major axis of ellipse equation derived by Hooghoudt has a value of half of _______.

a. Depth of impermeable layer from bottom of drain
b. Spacing between drains
c. Depth of impermeable layer from ground surface
d. None of these

92. The table of equivalent depths was developed by _______ by comparing the flows obtained with radial flow assumption and horizontal flow equation.

a. Wesseling
b. Hooghoudt
c. Ernst
d. Kirkham

93. The equivalent depths as per Hooghoudt's table are correct to about _______.

a. 1%
b. 10%
c. 2%
d. 5%

94. Artesian seepage is the upward movement of water from _______, which can add to the amount of water to be drained.

a. Unconfined aquifer
b. Confined aquifer
c. Semi-confined aquifer
d. All above

95. Leaching requirement is the _______.

a. Ratio of depth of drainage water to the depth of irrigation water
b. Ratio of depth of irrigation water to the depth of drainage water
c. Depth of drainage water
d. Depth of irrigation water

96. If the electrical conductivities of drainage water and irrigation water are 8 and 2 dS/m, respectively; the leaching requirement would be around _______.

a. 66% b. 25%
c. 40% d. 33%

97. The effect of the soil salinity is negligible on most of the crops when the EC of the soil saturation extract ranges between _______.

a. 2-4 dS/m at 25°C b. 4-8 dS/m at 25°C
c. 0-2 dS/m at 25°C d. None of these

98. In salt affected soils, only salt tolerant crops can yield satisfactorily when the EC of the soil saturation extract ranges between _______.

a. 8-16 dS/m at 25°C b. 16-20 dS/m at 25°C
c. 4-8 dS/m at 25°C d. None of these

99. In salt affected soils, the yields of many crops are restricted when the EC of the soil saturation extract ranges between _______.

a. 0-2 dS/m at 25°C b. 2-4 dS/m at 25°C
c. 4-8 dS/m at 25°C d. None of these

100. In salt affected soils, only a few very tolerant crops yield satisfactorily when the EC of the soil saturation extract is _______.

a. Above 8 dS/m at 25°C b. Above 16 dS/m at 25°C
c. Above 4 dS/m at 25°C d. Above 12 dS/m at 25°C

101. According to the Bureau of Reclamation, the solution to transient state of drainage is based on the assumption that the water table initially has a shape of __________.

a. Third degree parabola b. Ellipse
c. Second degree parabola d. Fourth degree parabola

102. As defined by the Bureau of Reclamation, the condition of annual recharge to be about equal to annual discharge with annual water table fluctuations becoming reasonably constant on yearly basis is _______.

a. Dynamic equilibrium b. Constant equilibrium
c. Intermittent equilibrium d. None of these

103. Not commonly used down wells in drainage engineering are also called _______.

a. Down draining
b. Drainage wells
c. Porous wells
d. All above

104. The use of down wells is limited to areas ___________.

a. Fractured lava
b. Porous limestone
c. Both (a) and (b)
d. None of these

105. The end-portion of the outlet pipe in subsurface drainage system should be blind (without perforations) and must at least be _______ long.

a. 0.3 m
b. 3 m
c. 1 m
d. Equal to drain width

106. While designing the subsurface drainage system, the outletpipe should be _______.

a. 60-100 cm above the maximum water level in the ditch
b. Equal to the depth of flow in ditch
c. 30-60 cm above the maximum water level in the ditch
d. Equal to the half of the depth of flow in ditch

107. The purpose of envelope material around subsurface drains is to __________.

a. Permit the fine clay particles to move through it
b. Prevent the fine and coarse sand to move through it
c. To support the drain
d. Both (a) and (b)

108. While designing a gravel envelope for uniform materials, the ratio of D_{50} sizes of filter and soil materials must range between _______ as suggested by the U.S. Bureau of Reclamation.

a. 5-10
b. 4-6
c. 10-12
d. 2-4

109. While designing a gravel envelope for graded materials, the ratio of D_{50} sizes of filter and soil materials must range between _______ as suggested by the U.S. Bureau of Reclamation.

a. 15-25
b. 12-58
c. 20-60
d. 6-12

110. Under ponded conditions, the flow into the drain is increased by _______ if we double the crack width between successive pipes in subsurface drainage system.

a. 20%　　b. 7%
c. 10%　　d. 11%

111. The use of _________ sections is feasible for the drains with envelope material without noticeable effect on flow into the drain.

a. Longer　　b. Shorter
c. Perforated　　d. Slotted

112. The recommended crack width between successive pipes in subsurface drainage system in stable soils is generally _______.

a. 0.30-0.45 mm　　b. 0.30-0.60 mm
c. 0.50-1.00 mm　　d. 0.45-0.60 mm

113. The increase in drain depth under ponded conditions ________ the amount of water entering the drain.

a. Increases　　b. Decreases
c. Does not affect　　d. None of these

114. With the increase in the number of holes in the drainpipe, the relative increase in inflow_______.

a. Increases　　b. Decreases
c. Is negligible　　d. None of these

115. The use of envelope material around the perforated pipe means that there will be _______ perforations necessary to get the desired flow.

a. Fewer　　b. More
c. No effect of the number of　　d. Chocking of

116. To determine the loads on a pipe placed in the bottom of a wide trench, _______ formula is used.

a. Ditch-conduit load　　b. Projecting conduit
c. Marston formula　　d. Both (b) and (c)

117. When trench is narrow, the load on a drainpipe is determined using the ______ formula.

a. Ditch conduit load
b. Projecting conduit
c. Marston formula
d. Both (a) and (c)

118. When trench width is 2-3 times the diameter of the pipe, the load on a drainpipe is determined using the ________.

a. Projecting-conduit formula
b. Wide trench formula
c. Marston formula
d. All above

119. The friction between fill material and walls of the trench _______ the load on the drain line due to weight of fill.

a. Decreases
b. Increases
c. Evenly distribute
d. Does not affect

120. To test the strength of rigid drainpipe with its bottom supported at two places, _______ test is used.

a. Sand-bearing
b. Three-edge bearing
c. Any of (a) or (b)
d. Crushing

121. To test the strength of a rigid drainpipe with the bottom part supported by a bed of sand, _______ test is used.

a. Sand bearing
b. Crushing
c. Three-edge bearing
d. Any of (b) or (c)

122. When subjected to the three edge-bearing test, a rigid drainpipe will support _______ load than when tested with the sand-bearing test.

a. Same
b. More
c. Less
d. None of these

123. The crushing strength of a rigid drainpipe tested with three-edge bearing test will be _______ times the crushing strength under the sand-bearing test.

a. 1.5
b. 1.0
c. 0.33
d. 0.66

124. To allow the surface drainage waters to percolate into the subsurface drainage system, _______ are used.

a. Blind inlets b. Surface inlets
c. By-pass lines d. Both (a) and (b)

125. For admitting surface water into buried drain, _______ is/are provided.

a. Blind inlets b. Surface inlet
c. By-pass lines d. Both (b) and (c)

126. For smallest wetted perimeter, cross-section of an open channel should be _______.

a. Trapezoidal b. Semi-circular
c. Rectangular d. Square

127. The side slopes of open ditch drain in sandy soils can be _______ or greater.

a. 3H:1V b. 2H:1V
c. 1H:3V d. 1H:2V

128. The side slopes of open ditch drain in clay and highly organic soils may range from _______ to almost vertical.

a. 2H:1V b. 1H:2V
c. 1H:1V d. 1H:0.5V

129. A water table well used for drainage is also called _______.

a. Gravity well b. Artesian well
c. Flowing well d. None of these

130. Gravity well is located in the _______ aquifer.

a. Confined b. Unconfined
c. Leaky d. Perched

131. The development of funnel shaped drawdown around the pumped well is known as _______.

a. Cone of influence b. Cone of impression
c. Cone of depression d. Both (a) and (b)

132. The removal of excess water from the surface of land by improved natural or constructed channels is called ________ system of surface drainage practice.

a. Diversion
b. Bedding
c. Random
d. Interceptor

133. The best-adapted drainage system for the areas with depressions too large or too deep to be filled by a land smoothing is _______ system.

a. Bedding
b. Interceptor
c. Random ditch
d. Diversion

134. The best surface drainage system for the poor internal drainage wet lands with slope of ______ or less is the cross-slope ditch system.

a. 1%
b. 4%
c. 10%
d. 5%

135. With the decrease in land slope, _______.

a. There is increase the spacing between cross-slope ditches
b. There is increase in the depth of cross-slope ditches
c. Both (a) and (b)
d. None of these

136. A _________ is that part of the drainage system, which receives water from the collection system and conveys it to an outlet.

a. Disposal conveyance system
b. Disposal system
c. Terminal point
d. Disposal outlet

137. A drainage system located across the groundwater flow direction is known as ________.

a. Relief drain
b. Across-slope drain
c. Interceptor drain
d. Seepage drain

138. The process of changing the surface of land to facilitate the movement of surface water over a field or a part of the field is known as __________.

a. Land forming
b. Land shaping
c. Land grading
d. All above

139. The shaping of land surface by cutting, filling, and smoothing to planned grades is known as ________.

a. Land shaping
b. Land grading
c. Land smoothing
d. None of these

140. The shaping of land surface to eliminate minor difference in elevation and to smooth out depressions without changing general contours of the land is known as _________.

a. Land grading
b. Land shaping
c. Land smoothing
d. All above

141. A ________ is a graded channel constructed across the land slope to intercept and divert water to a suitable outlet.

a. Diversion ditch
b. Cross-slope ditch
c. Interceptor ditch
d. Both (a) and (b)

142. The drains oriented at right angles to the direction of groundwater flow are called _______ drains.

a. Relief drain
b. Interceptor drain
c. Seepage drain
d. Perpendicular drain

143. The drains oriented parallel to the direction of groundwater flow are called _______ drains.

a. Seepage drain
b. Parallel drain
c. Relief
d. Interceptor drain

144. Generally, the drainage properties of ____________ are poor.

a. Platy soil structure
b. Alluvial plains
c. Heavy soils
d. Both (a) and (c)

145. Usually, the drainage coefficient value will be ________.

a. More for surface drainage than subsurface
b. Less for surface drainage than subsurface
c. Same for surface and subsurface drainage
d. None of these

146. Generally, a diversion channel is designed for a storm of ______ return period.

a. 5 years
b. 10 years
c. 25 years
d. 15 years

147. A system of parallel field ditches with intervening lands shaped to a convex surface resembles with the ______ system of drainage.

a. Diversion ditch
b. Random ditch
c. Bedding
d. Parallel ditch

148. The construction of beds in bedding system of drainage is practicable on land with a slope up to ______.

a. 2.5%
b. 0.5%
c. 1.0%
d. 1.5%

149. With increase in the permeability of soil, the ________ in bedding system.

a. Width of the bed increases
b. Slope of the bed increases
c. Both (a) and (c)
d. None of these

150. The capacity of a porous medium to transmit fluids when described in qualitative terms is known as ______.

a. Permeability
b. Porosity
c. Hydraulic conductivity
d. All above

151. The herringbone type of drainage system is a ______ type of drainage.

a. Interceptor
b. Seepage
c. Relief
d. None of these

152. The parallel system of relief subsurface drains is also called ______.

a. Herringbone system
b. Gridiron system
c. Relief system
d. Interceptor system

153. A ______ system is the modified form of the double main system of subsurface drains.

a. Gridiron
b. Parallel
c. Herringbone
d. Cross-slope

154. The diameter of mole drain generally _______.

a. Ranges between 7.5-10 cm
b. Ranges between 5.0-7.5 cm
c. Ranges between 10-12.5 cm
d. More than 12.5 cm

155. A_______ ditch or drain develops similar drawdown curves on its either side.

a. Seepage
b. Relief
c. Interceptor
d. Cut-off

156. For a perfect barrier, its permeability must be less than _______ of that of the overlying material.

a. 10%
b. 5%
c. 25%
d. 50%

157. The Hooghoudt's equation for drain spacing is also named as _______.

a. Ellipse equation
b. Two-dimensional equation
c. Donnan equation
d. All above

158. The Hooghoudt's equation for drain spacing is applicable where the flow of groundwater is largely _______.

a. Vertical
b. Sloppy
c. Horizontal
d. Radial

159. When depth to the barrier is less than _______ times the drain depth, the ellipse equation can be satisfactorily applied.

a. 5
b. 2
c. 1.5
d. 1

160. When all the groundwater flow enters the drain from the upslope side, it is called _______ drain.

a. Interception
b. Relief
c. Cut-off
d. Seepage

Key to objective type questions based upon chapters 1 to 4

Q. No.	Answer	Q. No.	Answer	Q. No.	Answer	Q. No.	Answer	Q. No.	Answer
1	(d)	41	(d)	81	(c)	121	(b)	161	(d)
2	(d)	42	(d)	82	(b)	122	(a)	162	(c)
3	(d)	43	(a)	83	(b)	123	(c)	163	(b)
4	(c)	44	(b)	84	(d)	124	(b)	164	(d)
5	(d)	45	(c)	85	(c)	125	(d)	165	(a)
6	(a)	46	(a)	86	(c)	126	(c)	166	(c)
7	(d)	47	(a)	87	(b)	127	(a)	167	(b)
8	(b)	48	(c)	88	(b)	128	(c)	168	(d)
9	(b)	49	(a)	89	(a)	129	(b)	169	(a)
10	(d)	50	(b)	90	(a)	130	(a)	170	(b)
11	(a)	51	(c)	91	(b)	131	(d)	171	(c)
12	(b)	52	(a)	92	(d)	132	(b)	172	(b)
13	(d)	53	(b)	93	(b)	133	(a)	173	(d)
14	(a)	54	(a)	94	(a)	134	(c)	174	(c)
15	(d)	55	(a)	95	(a)	135	(c)	175	(b)
16	(b)	56	(b)	96	(b)	136	(a)	176	(a)
17	(d)	57	(a)	97	(d)	137	(d)	177	(a)
18	(d)	58	(a)	98	(a)	138	(b)	178	(b)
19	(b)	59	(a)	99	(d)	139	(b)	179	(c)
20	(c)	60	(c)	100	(c)	140	(c)	180	(d)
21	(a)	61	(c)	101	(d)	141	(a)	181	(b)

Q. No.	Answer	Q. No.	Answer	Q. No.	Answer	Q. No.	Answer	Q. No.	Answer
22	(d)	62	(a)	102	(b)	142	(b)	182	(c)
23	(d)	63	(a)	103	(a)	143	(a)	183	(a)
24	(d)	64	(a)	104	(a)	144	(a)	184	(b)
25	(a)	65	(a)	105	(d)	145	(d)	185	(d)
26	(c)	66	(b)	106	(d)	146	(c)	186	(b)
27	(a)	67	(a)	107	(a)	147	(d)	187	(c)
28	(a)	68	(a)	108	(d)	148	(a)	188	(a)
29	(a)	69	(a)	109	(a)	149	(b)	189	(a)
30	(d)	70	(c)	110	(a)	150	(b)	190	(c)
31	(c)	71	(b)	111	(a)	151	(d)	191	(c)
32	(a)	72	(d)	112	(b)	152	(a)	192	(a)
33	(a)	73	(b)	113	(b)	153	(c)	193	(d)
34	(a)	74	(a)	114	(c)	154	(c)	194	(a)
35	(b)	75	(a)	115	(b)	155	(b)	195	(b)
36	(a)	76	(a)	116	(c)	156	(a)	196	(d)
37	(d)	77	(a)	117	(a)	157	(d)	197	(a)
38	(d)	78	(d)	118	(d)	158	(b)		
39	(d)	79	(c)	119	(a)	159	(a)		
40	(b)	80	(b)	120	(b)	160	(c)		

Key to objective type questions based upon chapters 5 to 11

Q. No.	Answer	Q. No.	Answer	Q. No.	Answer	Q. No.	Answer
1	(d)	41	(c)	81	(b)	121	(a)
2	(b)	42	(b)	82	(c)	122	(c)
3	(a)	43	(a)	83	(a)	123	(d)
4	(a)	44	(a)	84	(b)	124	(a)
5	(b)	45	(b)	85	(a)	125	(b)
6	(d)	46	(c)	86	(a)	126	(b)
7	(a)	47	(d)	87	(b)	127	(a)
8	(b)	48	(a)	88	(a)	128	(c)
9	(a)	49	(b)	89	(b)	129	(a)
10	(a)	50	(c)	90	(c)	130	(b)
11	(c)	51	(a)	91	(b)	131	(c)
12	(a)	52	(b)	92	(b)	132	(a)
13	(c)	53	(c)	93	(d)	133	(c)
14	(d)	54	(a)	94	(c)	134	(b)
15	(a)	55	(c)	95	(a)	135	(a)
16	(d)	56	(d)	96	(b)	136	(b)
17	(a)	57	(d)	97	(c)	137	(c)
18	(d)	58	(a)	98	(a)	138	(a)
19	(d)	59	(b)	99	(c)	139	(b)
20	(a)	60	(c)	100	(b)	140	(c)
21	(a)	61	(a)	101	(d)	141	(a)
22	(c)	62	(b)	102	(a)	142	(b)
23	(a)	63	(c)	103	(a)	143	(c)
24	(d)	64	(a)	104	(c)	144	(d)
25	(a)	65	(b)	105	(b)	145	(a)
26	(b)	66	(d)	106	(c)	146	(b)
27	(a)	67	(a)	107	(d)	147	(c)
28	(c)	68	(b)	108	(a)	148	(d)
29	(c)	69	(c)	109	(b)	149	(a)
30	(d)	70	(b)	110	(c)	150	(a)
31	(b)	71	(a)	111	(a)	151	(c)
32	(b)	72	(c)	112	(b)	152	(b)
33	(a)	73	(d)	113	(a)	153	(c)

34	(b)	74	(a)	114	(b)	154	(a)
35	(c)	75	(b)	115	(a)	155	(b)
36	(b)	76	(d)	116	(d)	156	(a)
37	(d)	77	(a)	117	(d)	157	(d)
38	(d)	78	(d)	118	(d)	158	(c)
39	(a)	79	(d)	119	(a)	159	(b)
40	(b)	80	(c)	120	(b)	160	(a)

Literature cited

Abrol, I. P., Yadav, J. S. P., and Massoud, F. I. (1988). Salt-affected soils and their management. Soils Bulletin 39. FAU, Rome. 131p.

Ahrens, T. P. (1958). Water well engineering, more on well design criteria. Water Well Journal, 12: 11-12.

Ali, G., Asghar, M. N., Latif, M., and Hussain, Z. (2004). Optimizing operational strategies of scavenger wells in lower Indus basin of Pakistan. Agricultural Water Management, 66: 239-249.

Anonymous. (2005) All India Coordinated Research Project on Groundwater Utilization. Department of Soil and Water Engineering, college of Agricultural Engineering and Technology, Punjab Agricultural University, Ludhiana, Punjab.

Aronovici, V. S. (1947). The mechanical analysis as an index of subsoil permeability. Proceedings of the American Soil Science Society, 11: 137-141.

ASAE. (2005). ASAE standards. The Society for engineering in agricultural, food, and biological systems. 52nd Edition. 1044p.

Bennison, E. W. (1947). Ground Water. Edward E. Johnson, St. Paul, Minnesota. 440p.

Bureau of Indian Standards. (1962). Indian standard specification for test sieves. IS 460: 1962

Bureau of Indian Standards. (1981). Specification for leaded brass strip for instrument parts. IS 531: 1981. New Delhi.

Bureau of Indian Standards. (2000). Well screens and slotted pipes – specification. IS 8110:2000. New Delhi.

Carman, P. C. (1937). Fluid flow through granular beds. Transactions – Institution of Chemical Engineeres, 15: 150-166.

CGWB. (2011). Dynamic groundwater resources of India. Central Ground Water Board, Ministry of Water Resources, River Development & Ganga Rejuvenation, Government of India. 282p.

Childs, E. C., and Collis-George, N. (1950). The permeability of porous materials. Proceedings of the Royal Society (London), A201: 392-405.

Chow, V. T. (1952). On the determination of transmissibility and storage coefficients from pumping test data. Transactions of American Geophysical Union, 33: 397-404.

Church, A. H. (1944). Centrifugal pumps and blowers. John Wiley & Sons, Inc., New York. 307p.

Cooper, H. H. Jr., and Jacob, C. E. (1946). A generalized graphical method for evaluating formation constants and summarizing well-field history. Transactions of American Geophysical Union, 27(4): 526-534.

Darcy, H. (1856). Les fontaines publiques de la ville de Dijon, V. Dalmont, Paris. 647p.

Freeze, R. A., and Witherspoon, P. A. (1966). Theoretical analysis of regional groundwater flow: 1. Analytical and numerical solutions to the mathematical model. Water Resources Research, 2(4): 641-656.

Frevert, R. K., Schwab, G. O., Edminster, T. W., and Barnes, K. K. (1955). Soil and water conservation engineering. Soil Science, 80(1), 89p.

Grainage, J. W., and Lund, E. (1969). Quick culturing and control of iron bacteria. Journal of American Water Works Association, 61(5): 242-245.

Hantush, M. S. (1964). Hydraulics of wells. Advances in hydroscience, 1: 281-432.

Haynes, H. D. (1966). Machinery and methods for constructing and maintaining surface drainage on farm lands in humid areas. Transactions of ASAE, 9(2): 185-189, 193.

ICID. (1982). Committee on irrigation and drainage construction techniques. ICID standard 109, Construction of surface drains. ICID Bulletin, 31(1): 47-57.

Johnson, E. E. (1966). Ground water and wells: a reference book for the water-well industry. Edward E. Johnson Inc., Saint Paul, Minnesota, USA. 440p.

Khepar, S. D., Kaushal, M. P., and Katyal, A. K. (1975). Development of an animal drawn lift irrigation pump. Journal of Agricultural Engineering, 12(3-4):42-47.

Kirkham, D. (1946). Proposed method for field measurement of permeability of soil below the water table. Proceedings of the American Soil Science Society, 10: 58-68.

Kirkham, D. (1958). Seepage of steady rainfall through soil into drains. Transactions of American Geophysical Union, 39: 892-908.

Kirkham, D., and Schwab, G. O. (1951). The effect of circular perforations on flow into subsurface drain tubes. Part I. Theory. Agricultural Engineering, 32(4): 211-214.

Kozeny, J. (1927). Ueber capillare Leitung des Wassers im Boden. Sitzungsber. Wien. Akad. Wissensch., 136 (2a): 271-306.

Linsley, R. K., and Franzini, J. B. (1979). Water-resources engineering. New York, McGraw-Hill.

Lohman, S. W. (1972). Ground-water hydraulics. Professional Paper No. 708. U.S. Geological Survey. 70p.

Luthin, J. N. (1973). Drainage Engineering. John Wiley & Sons Inc. 250p.

Luthin, J. N. and Haig, A. (1972). Some factors affecting flow into drain pipes. The National Academies of Sciences, Engineering, and Medicine, 41(10): 235-245.

Marshall, T. J. (1957). Permeability and the size distribution of pores. Nature, 180: 664-665.

Marston, A. (1930). The theory of external loads on closed conduits in the light of the latest experiments. Bulletin of Iowa Engineering Experiment Station, 96p.

Mehta, K. K. (1983). Reclamation of alkali soils in India. Oxford and IBH Publishing Co., New Delhi. 280p.

Michael, A. M. (1997). Irrigation, theory and practice. Vikas Publishing House Pvt. Ltd. New Delhi. 801p.

Mogg, J. L. (1972). Practical corrosion and incrustation guidelines for water wells. Groundwater, 10(2): 6-11.

Mohammad, F. S., and Skaggs, R. W. (1983). Drain tube opening effects on drain inflow. Journal of Irrigation and Drainage Engineering, 109(4): 393-404.

Ritzema, H. P. (1994). Drainage principles and applications. ILRI publication 16 (second edition), The Netherlands. 1125p.

Rorabaugh, M. I. (1953). Graphical and theoretical analysis of step-drawdown test of artesian well. Proceedings of American Society of Civil Engineers, 79 (Sep. 362). 23p.

Singh, S. R., and Shakya, S. K. (1989). A nonlinear equation for groundwater entry into well screens. Journal of Hydrology, 109(1): 95-114.

Spencer, W. F., Patrick, R., Ford, H. W. (1963). The occurrence and cause of iron oxide deposits in tile drains. Proceedings of the American Soil Science Society, 27: 134-137.

Theis, C. V. (1935). The relation between the lowering of the piezometric surface and the rate and duration of discharge of a well using ground-water storage. Transactions of American Geophysical Union, 16: 519-524.

Vashisht, A. K., and Shakya, S. K. (2015). Drainage of inundated land through a well partially penetrating the single leaky aquifer. Proceedings of the Hydro 2015 International which was held at IIT Roorkee on 17-19 December 2015. Organized by Indian Society for Hydraulics, Pune, India.

Vashisht, A. K., and Shakya, S. K. (2016a). Development of semi-analytical solution for recharging through multiple well points system under constant head conditions. ISH Journal of Hydraulic Engineering (ASCE), 22(1): 50-58.

Vashisht, A. K., and Shakya, S. K. (2016b). Flow through multiple well points system. In Groundwater: Contaminant and Resource Management, InTech Publishers, Croatia. 45-68pp.

Walker, W. H. (1974). Tube wells, open wells, and optimum ground-water resource development. Groundwater, 12(1): 10-15.

Wurts, W. A., McNeill, S. G., and Overhults, D. G. (1994). Performance and design characteristics of airlift pumps for field applications. World Aquaculture, 25(4): 51-55.

W